THE ORIGIN OF SUBMARINE CANYONS

NUMBER III OF THE
COLUMBIA GEOMORPHIC STUDIES

EDITED BY

DOUGLAS JOHNSON

PLATE I: Blue Lakes Alcove, Idaho. A canyon produced by solution spring sapping in basaltic lavas.

THE ORIGIN OF SUBMARINE CANYONS

A CRITICAL REVIEW OF HYPOTHESES

BY

DOUGLAS JOHNSON

PROFESSOR OF PHYSIOGRAPHY
COLUMBIA UNIVERSITY

NEW YORK
COLUMBIA UNIVERSITY PRESS
1939

Copyright 1939
COLUMBIA UNIVERSITY PRESS
Published 1939

Foreign Agents

OXFORD UNIVERSITY PRESS
Humphrey Milford, Amen House
London, E.C.4, England

MARUZEN COMPANY, LTD.
6 Nihonbashi, Tori-Nichome
Tokyo, Japan

OXFORD UNIVERSITY
B. I. Building, Nicol Road
Bombay, India

Printed in the United States of America

Preface

Few problems in the field of geomorphology have proven so baffling as that of finding a satisfactory explanation for the deeply submerged canyons indenting the seaward margins of many continental shelves. For over half a century the problem has been debated, and with increasing intensity during the last decade. Yet there is today no concensus of opinion as to how these remarkable features were produced.

During the course of debate many misstatements of fact and many misinterpretations of opinion have crept into the voluminous literature published on the question. Such errors inevitably complicate a problem already sufficiently complex, and render more difficult discovery of the correct solution. It therefore seemed to the writer that he might render a useful service to geomorphology by preparing a critical review of what has thus far been accomplished in the search for an explanation of submarine canyons. Such is the primary purpose for which the present work was undertaken.

As the work progressed a secondary purpose increased in relative importance. Studies of the Carolina "bays" (explained by Melton and Schriever as oval depressions produced by a shower of giant meteorites striking the earth at an oblique angle) convinced the writer that retrogressive artesian spring sapping had played an important rôle in forming the elongated "craters." From this conception sprang the hypothesis that submarine canyons might have a somewhat similar origin. If the step from Carolina bays to submarine canyons seems both long and illogical, the answer is that this particular association of ideas was not the result of logical reasoning. It came unbidden and instantaneously—which is perhaps the way most hypotheses are "invented."

Once the hypothesis was in mind, it seemed worth while to explore its possibilities. Such exploration was the more necessary because in the futile search for a satisfactory explanation of submarine canyons most investigators had confined themselves chiefly to consideration of subaërial and submarine agencies. This attitude is well exemplified by Stetson, who concludes one of his valuable discussions with the remark: "At the present writing, the question of subaërial versus submarine origin of the canyons is regarded as still open. The evidence at hand does not warrant a complete rejection of either view." On general principles it seemed most likely that a solution for the problem might be found where it had least been sought: in the subterranean realm.

Elaboration and testing of the hypothesis of artesian spring sapping assumed ever-increasing importance, until results of this phase of the study came to constitute a significant portion of the present volume. At its inception the hypothesis seemed little more than a wild flight of fancy. On analysis it appears to offer more hope for a satisfactory solution of the submarine canyon problem than does any one of the many explanations previously advanced.

Most of the material in this volume appeared as a series of articles in the *Journal of Geomorphology*. In preparing the text for publication in book form the writer has taken advantage of the opportunity to revise in certain respects the manner of presentation, to add some new observations, citations, and illustrations, and to revise the French résumés in the form of a single summary chapter.

Both the work of revision and the preparation of the original manuscript would have been practically beyond the writer's power but for the loyal and efficient services of his research assistants Mrs. Girard Wheeler and Miss Margaret Cooper, and his secretary Mrs. John Torpats. For the heavy labors they performed, and for the generosity of Columbia University in making their services available, the writer owes his greatest debt of gratitude.

In preparing this study the writer has appealed freely to so many colleagues that full acknowledgment is impossible. To colleagues in the Department of Romance Languages at Columbia he is indebted for revision of his efforts to put the substance of the work into French. Among scientific colleagues to whom he is chiefly obligated for valuable suggestions and criticisms are O. E. Meinzer, C. Wythe Cooke, J. S. Brown, C. P. Berkey, S. J. Shand, G. M. Kay, H. S. Sharp, A. D. Howard, Girard Wheeler, J. H. Mackin, R. F. Flint, M. King Hubbert, and L. W. Stephenson. It is understood, of course, that acknowledgment of indebtedness to these gentlemen implies no approval by them of views expressed in the following pages.

DOUGLAS JOHNSON

Columbia University
September 15, 1939

Contents

Chapter I. Introduction 3
   1. HYPOTHESES INVOLVING A TECTONIC ORIGIN 4
   2. HYPOTHESES INVOLVING A NON-TECTONIC ORIGIN . . 5

Chapter II. Hypotheses of Subaërial Origin 8
   1. RECENTLY SUBMERGED RIVER GORGES 8
   2. ANCIENT SUBMERGED RIVER GORGES RE-EXCAVATED BY LAND-SLIDING 9
      Evidence of Submarine Landslides 12
   3. GROUNDWATER SAPPING 18

Chapter III. Hypotheses of Submarine Origin 23
   1. SUBMARINE LANDSLIDES 23
   2. SUBMARINE CURRENTS 24
      Various Types of Currents 24
      Turbidity Currents 27
      Strength and Weakness of the Turbidity Hypothesis . 33
      Sublacustrine Trenches of the Rhine and Rhone Rivers 35
      Currents in the Strait of Gibraltar 44
      Effect of Salinity on Deposition of Fine Sediment . . 47
      Degree of Lithification of Shelf Sediments . . . 52
      Effects of Salinity and Temperature on Turbidity Currents in the Ocean 55
      Inadequacy of Turbidity Currents 57
      Flow of Turbid Water through Reservoirs 59
      Conclusion Respecting Turbidity Currents 61

VIII *Contents*

Chapter IV. Hypotheses of Subterranean Origin 66
   1. SUBTERRANEAN RIVER OUTLETS 66
   2. FOUNDERING OF SUBTERRANEAN CAVERNS 67
   3. SOLUTION ALONG FAULTS BY UP-RISING SUBTERRANEAN WATERS 68
   4. SOLUTION ALONG FAULTS BY DOWN-FILTERING MARINE WATERS 68
   5. NON-DEPOSITION ALONG FAULTS DUE TO UP-RISING SUBTERRANEAN WATERS 69
   6. SUBMARINE SPRING SAPPING 71
       A. Spring Development on the Continental Slope . 72
           Submarine Springs in Shallow and Deep Water 72
           Artesian Conditions in the Continental Shelves 74
           Artesian Conditions in the Geologic Past . . 82
           Conditions Affecting the Outcrop of Artesian Horizons 86
           Summary Respecting Artesian Conditions . . 88
           Expulsion of Non-artesian Waters from Continental Shelves 89
       B. Competence of Submarine Springs to Excavate Canyons 91
           Rôle of Solution in Spring Sapping 93
           Rôle of Mudflowing 103
       C. Form and Distribution of Canyons Compatible with Spring-sapping Hypothesis 105
       Conclusion 107

Résumé (in French) 113

Index 121

Illustrations

PLATES

I. Blue Lakes Alcove, Idaho . . . . *Frontispiece*

II. The Hudson Submarine Canyon 10

III. Submarine Canyons along Southern Margin of Georges Bank   18

IV. Submarine Canyon and Furrowed Outer Slope of Continental Shelf Southeast of Delaware Bay 50

FIGURES

1. Diagram Showing Artesian Conditions Producing Submarine Springs 76

2. Artesian Conditions Permitting Leakage out of or into an Aquifer 80

3. Portion of the Older Appalachians of Northern New Jersey, Coastal Plain Remnant, and Probable Former Continuation of the Coastal Plain Beds 84

4. "Steep-Head" Valleys Produced by Spring Sapping in the Coastal Plain of Northwestern Florida 92

THE ORIGIN OF SUBMARINE CANYONS

CHAPTER I

Introduction

THE OUTER MARGINS of continental shelves are frequently notched by gigantic gullies or ravines popularly called "submarine canyons." These deeply submerged trenches commonly measure from one to several miles across at the top and are cut from one to several thousand feet below the adjacent surface of the shelf, and sometimes as much as eight or ten thousand feet, possibly more, below sea level. They are cut back into the shelf for varying distances, five to ten miles being common while a few examples must be measured in scores of miles. Some of these marginal notches appear to connect with much longer but comparatively shallow channels crossing the shelf from the vicinity of the shore. Some are opposite the mouths of large rivers. But many appear unrelated to channels in the shelf or to valleys in the exposed lands.

One should not confuse the deeply submerged forms above described with the shallower "drowned river valleys" indenting the coasts of many lands, with the deeper fjords indenting glaciated coasts of high latitudes and extending far out beneath the sea, or with the comparatively shallow submarine channels with which they sometimes appear to connect. That the "canyons" notching the shelf margins may have the same origin as the forms just named is a possibility to be duly considered. But until identity of origin is established the deeply submerged shelf-margin notches, or "submarine canyons," are better classed by themselves. It is with them alone that the present volume is concerned.

Submarine canyons of the type here considered began to attract attention more than half a century ago, but it is only within recent years that their forms have been determined with any approach to accuracy, thanks to improved methods of sounding the ocean depths. Fifty years of discussion, and much new knowledge of the characteristics of the canyons, have not, however, brought agreement among geologists respecting the genesis of these forms. Many hypotheses of origin have been advanced, but none has yet attained the standing of a generally accepted theory.

One must keep in mind the possibility that there is no single explanation for the canyons. As our knowledge of submarine topography is increased by more abundant soundings and by other means, it may be found that somewhat similar yet genetically distinct forms are now being wrongly classed together under the term "submarine canyons." The geomorphologist of the future may be in possession of facts enabling him to divide submarine canyons into several unlike groups, and to assign to each a different origin, just as today we class apart and explain differently the drowned normal valleys of middle and low latitudes and the partially submerged glacial troughs of high latitudes. But in the present imperfect state of our knowledge we can only say that while occasional canyons, such as that off the mouth of the Congo and the famous "fosse de Cap-Breton" under the Bay of Biscay, show certain peculiarities which may set them apart, the apparent similarity of form of most of the canyons, the similarity of their occurrence as notches in the submerged margins of the continents, and the degree of similarity of their depths below the shelf surface and below sea level do not justify distinctions between them. Accordingly one is impelled to seek a common origin for that great majority of the canyons which at present appear to exhibit common characteristics. It seems pertinent, therefore, to review more fully than has yet been done the various hypotheses of canyon origin already advanced, and to offer a new hypothesis which seems worthy of critical examination. It is to the former task that attention is first directed.

I. HYPOTHESES INVOLVING A TECTONIC ORIGIN

In seeking to explain submarine canyons some authors have appealed to types of faulting, warping, or folding which would produce long and relatively narrow troughs. Such structures could be either formed under the ocean, or formed above sea level and later submerged. Faulting might occur along a single plane with tilting producing a fault-trough, or an elongated block between parallel faults might drop to give a graben. Warping could produce a broadly open synclinal basin, or sharper folding a deep synclinal trough.[1]

Structural interpretations of the types indicated were, for the most part, offered before the detailed forms of submarine canyons became as fully known as at present. Some authors have grouped in a single class the shelf-margin canyons here considered and deep oceanic troughs beyond the shelf or between oceanic islands, invoking a common origin

Non-tectonic Origin

for all. While a tectonic origin for the second group of forms can be sustained with weighty arguments, and while an occasional canyon in the shelf margin may be due directly or indirectly to tectonic causes, few geologists would today offer such an explanation for the great group of shelf-margin canyons.

Recent studies of typical examples of this latter group have brought to light new and significant facts which are not readily explained on a structural basis. The major canyons frequently exhibit more or less sinuous courses, and usually descend seaward with fair continuity of bottom slope. Tributary side canyons sometimes occur, and are arranged in a somewhat dendritic pattern. Both major and tributary canyons most frequently have V-shaped cross-profiles. The canyons are bordered on either side, and landward from their heads, by a moderately smooth continental shelf which shows no indication of having suffered major yet sharply localized structural disturbance.

The difficulties of applying any structural interpretation to such forms is so evident, and abandonment of hypotheses of tectonic origin is today so nearly universal, that one need not further elaborate this phase of the subject.

2. HYPOTHESES INVOLVING A NON-TECTONIC ORIGIN

The remainder of this volume is devoted to a consideration of those hypotheses of submarine canyon development which do not involve a tectonic origin. Hypotheses of non-tectonic origin are very numerous, if we pay respect to minor differences existing between some of them. But for purposes of discussion all may be treated under less than a dozen headings which fall naturally into three major groups: hypotheses of subaërial origin, hypotheses of submarine origin, and hypotheses of subterranean origin.

The first group, hypotheses of subaërial origin, comprises three distinct interpretations: first, that submarine canyons are normal river gorges formed above sea level and recently submerged; second, that they are ancient subaërial river gorges which were filled with debris and later re-excavated by landsliding; and third, that they are the product of subaërial groundwater sapping.

Under the second group, hypotheses of submarine origin, brief consideration will be given to the little-supported hypothesis that the canyons resulted originally and directly from submarine landsliding. Attention will be directed chiefly to various phases of the hypothesis that suboceanic canyons have been carved by some type of submarine current. Because of the increasing attention recently given to turbidity currents (sometimes called "suspension currents") as a possible cause of submarine canyon cutting, the character of such currents and their competence as erosive agents will be examined at length.

The third group, hypotheses of subterranean origin, consists of half a dozen interpretations in which the cause of the canyons is sought in conditions existing within those parts of continental masses projecting beneath the ocean. In a sense, canyon formation according to these hypotheses is both subterranean and submarine. But since the controlling conditions are believed to lie within the earth's crust, it seems permissible, and certainly is convenient, to distinguish this group of hypotheses from the group called "submarine" above, by referring to them simply as "hypotheses of subterranean origin." Included in this group are the following explanations: that the canyons represent the outlets of subterranean rivers; that they were caused by foundering of the roofs of subterranean caverns; that they were produced along faults by the solvent action of up-rising subterranean waters; that they were produced under similar geological conditions, but by down-filtering marine waters; that they result from non-deposition along faults where up-rising subterranean waters prevented the accumulation of sediments; and finally, the hypothesis here presented for the first time, that the canyons have been excavated through the sapping action of springs of artesian and other waters issuing far down the seaward faces of continental shelves.

In the following chapters the various hypotheses enumerated above will be treated in the order indicated.

Notes and References

1. Among comparatively recent papers which suggest the possibility of a structural origin, in whole or in part, for features classed as submarine canyons, the following may be cited:

A. C. Lawson, "The Geology of Carmelo Bay." *Univ. of Calif., Bull. Dept. Geol.*, Vol. 1, pp. 1-59, 1893.

———."The Continental Shelf off the Coast of California." *Bull. Nat. Research Council*, Vol. 8, Part 2, No. 44, 23 pp., 1924.

A. W. Wegener, *The Origin of Continents and Oceans*. (London) 212 pp., 1922. See p. 178.

Naomasa Yamasaki, "Physiographical Studies of the Great Earthquake of the Kwanto

District, 1923." *Jour. Fac. Sci., Imperial Univ. of Tokyo*, Vol. 2, Sect. 2, pp. 77-119, 1926.

J. W. Gregory, "The Earthquake South of Newfoundland and Submarine Canyons." *Nature*, Vol. 124, pp. 945-946, 1929.

———"The Earthquake off the Newfoundland Banks of 18 November 1929." *Geog. Jour.*, Vol. 77, pp. 123-134, 1931.

———"A Submarine Trough off the Coast of Cyprus." *Geog. Jour.*, Vol. 78, pp. 357-361, 1931.

———"A Submarine Trough near the Strait of Gibraltar." *Geog. Jour.*, Vol. 79, pp. 219-220, 1932.

CHAPTER II

Hypotheses of Subaërial Origin

1. RECENTLY SUBMERGED RIVER GORGES

The most obvious explanation of submarine canyons offered more than fifty years ago and still regarded by many investigators as the one most probably correct, is that they are normal young river valleys or gorges, carved during higher stands of the continents or a lower stand of sea level, and deeply submerged in comparatively recent geologic time. The strongest support for this interpretation of canyon history is found in the close resemblance of the submarine forms to stream gorges now observable on the lands. Shepard, in the series of articles cited under his name below and in a later section, presents most fully the arguments in favor of the submerged river gorge hypothesis;[1] while the objections to it have recently been effectively summarized by Daly.[2]

Despite the simplicity of this hypothesis, and its apparent validity in explaining relatively shallow drowned valleys indenting the lands and shallow submarine trenches on the surface of the continental shelf, it has not secured general acceptance as an explanation of shelf-margin canyons. When applied to canyons cut thousands of feet deep, the hypothesis calls for recent vertical oscillations of land or sea of very great magnitude. There are obvious theoretical objections to the conception that margins of continental shelves in most parts of the world have been raised many thousands of feet out of the sea, held there for a brief period only, then lowered approximately to their former level, all within a comparatively short lapse of time geologically speaking. Neither continental shelves nor adjacent lands afford independent evidence of such vast diastrophic disturbances.

The alternative conception, that sea level dropped many thousands of feet and then rose again, due to glaciation and deglaciation of the continents, to successive changes in the ocean floor opposite in sense but equal in magnitude, or to other causes, seems equally difficult of adoption. Glacial lowerings of sea level are generally believed to be of the order of magnitude of 300 feet. The lowering of "3,000 feet or more" recently invoked by Shepard,[3] while ten times that usually attributed to this cause and in the light of our limited knowledge apparently highly excessive, does not touch the problem of canyons cut 5,000 to 10,000 feet below present sea level. Even his earlier estimate of 6,000 feet leaves the deeper canyons unexplained. Neither ocean bottom changes nor other terrestrial causes adequate to produce such extensive variations of sea level are known to geologic science. Appeals to astronomical causes,[4] while deserving hospitable consideration, carry one so far into the realm of speculative reasoning as to offer little solid foundation for a satisfying explanation of such sea level changes. The coastal borders of the lands do not show evidence of trenching appropriate to any important relative lowering of baselevel.

Assuredly there are weighty objections to that hypothesis of submarine canyon origin which on its face seems most obvious and most simple. It is not surprising, therefore, that in recent years there has been an increasing tendency for students of these puzzling features to seek some other explanation which, while less obvious, shall encounter less formidable difficulties. As we shall discover, the search for an alternative explanation has not been sufficiently successful to displace definitely the hypothesis that the canyons are recently submerged river gorges. This still remains a working hypothesis of which account must be taken by all students of the problem.

2. ANCIENT SUBMERGED RIVER GORGES RE-EXCAVATED BY LANDSLIDING

In 1931 Shepard[5] expressed the opinion that submarine canyons on the south side of Georges Bank were cut when the continental shelves were elevated above the sea, thus making the shelves older than the canyons. A year later he wrote that the canyons of this region were cut and "submerged before the development of the present continental shelf, perhaps many millions of years ago."[6] This latter view was maintained in 1933, the time of canyon cutting being placed tentatively "well back in the geological time scale" in one paper, and assigned in part to the Tertiary in another.[7] The following year the canyon cutting was again placed at "remote" periods, not all necessarily at the same time; and it was stated that cutting might have occurred as far back as the late Palaeozoic.[8] Shepard was still convinced that canyon cutting preceded development of the continental shelves, his idea of the whole sequence of events being best stated in the following quotation, taken from the paper last cited: "The uplift which allowed the canyon cutting was clearly prior to the formation of the continental shelf in this vicinity [off the New England coast], and may have been as remote as the late Paleozoic. The canyon walls are almost certainly solid rock, indicating that the continental shelf here has been the product of erosion rather than of building up by sediments from the lands. The canyons are thought to have been filled with sediment in whole or in part, but are preserved in outline, by the evacuation of the sediment through the agency of landslides." The more common interpretation, that the canyons were first cut and submerged in relatively recent geologic time, had previously been definitely rejected; and the view was expressed that lowering of sea level by the formation of the great ice sheets was "not nearly enough [only 300 or 400 feet] to account for the [submarine] valleys," and hence "scarcely merits much consideration" as an explanation of the occurrence of these forms.[9]

The hypothesis of ancient drowned gorges, by making the valleys older than the adjacent continental shelf, and by requiring that shelf to be in large part a marine erosional rather than a depositional feature, appears to create more difficulties than it solves. Available evidence strongly negatives the idea that marine erosion has been a major factor in the development of our present continental shelves.[10] The form of the shelf bordering eastern North America, the materials dredged from it, and recent geophysical researches all indicate that it is the seaward continuation of Cretaceous, Tertiary, and later coastal plain deposits. Fossiliferous material from walls of deeply submerged marginal canyons seems to show that at least some of these canyons were cut in late Tertiary or Pleistocene time.

In 1935 Shepard published two papers, in the first[11] of which he still supported the ancient drowned valley hypothesis, whereas in the second[12] he seemingly abandoned it. Instead of postulating continental uplift and erosion of valleys in remote geological time, followed by submergence in a sea which changed its level but little, the second of these papers held that "The accumulated evidence in regard to submarine canyons has made it difficult to avoid the conclusion that these canyons were cut as a result of a lowering of sea level of at least 6,000 feet in not very remote times." The following year Shepard placed the cutting of some of the canyons in "very late Tertiary or Pleistocene

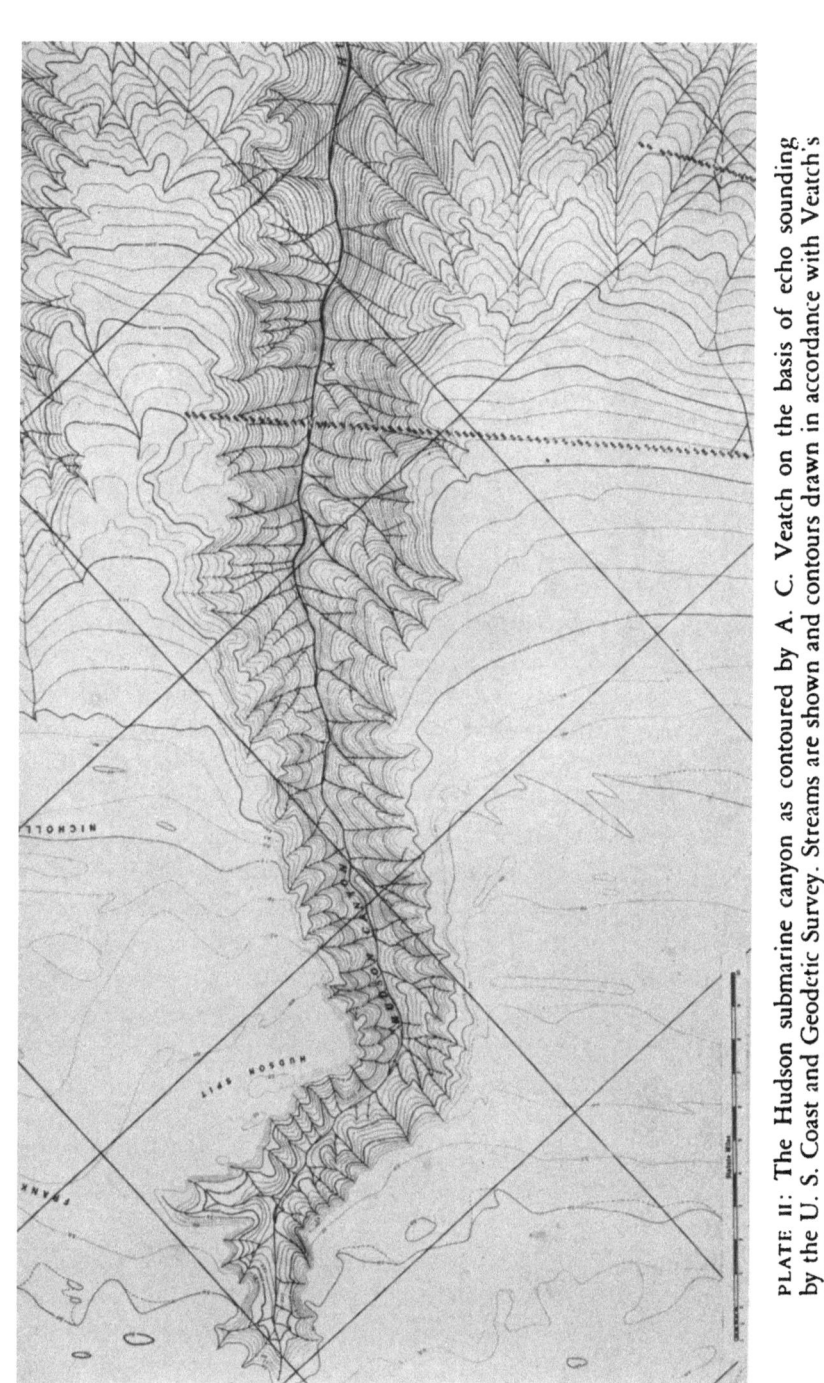

PLATE II: The Hudson submarine canyon as contoured by A. C. Veatch on the basis of echo sounding by the U. S. Coast and Geodetic Survey. Streams are shown and contours drawn in accordance with Veatch's belief that submarine canyons represent submerged subaërial valleys. Contour interval: on shelf, 5 fathoms (30 feet) to depth of 85 fathoms; on slope 25 fathoms (150 feet). Reproduced from Chart IB, Special Paper No. 7, Geological Society of America.

Re-excavation by Landsliding

.... very likely mid-Pleistocene";[13] and a year later still others were believed to be carved "during a great drop in sea level," presumably the extreme glacial lowering which he had recently begun to advocate.[14] Shortly prior to publication of the paper last cited, in a discussion of "the underlying causes of submarine canyons," Shepard apparently endeavored to harmonize his earlier and later views in a composite explanation embodying elements of both. He writes: "The following suggestions are the best that the writer can offer at present as a general explanation: First, that prior to the glacial period there were various depressions on the continental slopes which were in part the result of landslides or mudflows, in part diastrophic, and in part true river canyons submerged by diastrophism. Second, that the sea level was lowered 3,000 feet or more during the maximum glaciation of some early epoch and that the rivers from the land flowed into the various preëxisting depressions and cut true river canyons through relatively recent sediments connecting these canyons with the deep outer depressions. Third, that the sea level rose with the melting of the ice and that the canyons have been maintained ever since on the steep slopes due to the mudflows, which are still occurring from time to time and that currents may have played some part in keeping the canyon heads clear of sediment."[15]

It is evident that Shepard has here formulated what is in effect an omnibus explanation in which the development of ancient stream-carved valleys, landslide depressions, and diastrophic (tectonic?) basins, extensive subsidence of the land, extensive glacial lowering of sea level, recent stream erosion, mudflows, and submarine currents shall all play a part. While it is quite possible that various processes of nature may have operated in producing submarine canyons, unless evidence of the part played by each is clearly recognizable any explanation which invokes so large a number of causes must remain too generalized to be satisfying.

Submarine landslides and mudflows are possible accompaniments of submarine canyon development under almost any theory of canyon origin. But under the hypothesis of submerged ancient river gorges they become an essential element, being required to clear the ancient gorges of accumulating sediments which must otherwise have filled and concealed them. Because the strength of this hypothesis of canyon origin depends in no small measure on proof that landslides or mudflows have been an active factor in clearing gorges previously wholly or partially filled with loose debris, it is pertinent to examine available evidence of such activity.

Evidence of Submarine Landslides. The occurrence of submarine landslides has sometimes been inferred from the breaking of cables, both in connection with submarine earthquakes, and without evidence of associated seismic activity. Occasionally fractured ends of the cables appear to have been buried under debris and torn or twisted, while in one case a cable was reported buried "for more than 100 miles." Milne[16] summarized some of the data available in 1897, while Gregory,[17] Shepard,[18] and others have reported on more recent examples.

It may be questioned whether the great interest attaching to any evidence of movement on the floor of the ocean has not led to some exaggeration of the part landslides play in the rupturing of cables. The breaking of a cable during an earthquake does not necessarily prove the occurrence of a submarine landslide. Sudden displacement of the ocean floor by faulting is alone sufficient to cause rupture. Such displacement may, of course, give rise to slumping along the fault scarp, and fractured cable ends may then be buried in landslide debris. But landslides of this type, assuming they occur, are due to the creation of an over-steepened cliff, and have nothing to do with the evacuation of filled submarine canyons by landslipping or mudflowing initiated by earth tremors. Nor is it safe to assume either slumping along a fault scarp, or other types of landslides, merely because the cables show evidence of having been buried and distorted. Sedimentation over a long period of years may have buried the cables for great distances; while rupture by displacement of the ocean floor may account for twisting or tearing of fractured cable ends.

It thus appears that while landslides may well occur on steep portions of the ocean floor, including the steep face of the continental slope, records of cable fracturing must be critically scrutinized before they are accepted as proof of such slides. In no instance can cable breaks as yet be definitely correlated with the sliding or flowing of debris accumulated as fill in older submarine valleys. Not until it is clearly demonstrated that cables passing along or across the courses of submarine canyons, or opposite their mouths, have been broken by masses of debris which could not have slumped from the face of fault scarps developed on the ocean floor at the time of seismic activity, or from the steep face of the continental slope, will evidence from cable breaks lend strength to the hypothesis of landslide evacuation of older submarine canyons.

Shepard has presented other evidence of landsliding or mudflowing from certain submarine canyons. Thus he notes[19] that Captain Bone of the steamship *Transylvania*, crossing the Georges Bank area shortly after the occurrence of the Grand Banks earthquake of November 18, 1929, reported deep water where the chart showed shallow depths on the Bank, and concluded that the earthquake had altered the sea bottom at this point. Later the Coast and Geodetic Survey, while resurveying the Bank, found in this vicinity a deep submarine valley cutting seven miles back into the shelf. Shepard states that the older soundings "were so far apart that the valley might easily have been missed." But he expresses the view, based on the fact that vessels coming in from Europe have been taking soundings in this general vicinity for years, that it is "a little strange that it was not till after the earthquake that the depression was suspected." A year later Shepard[20] cites this case under the subheading "Recent Submarine Landslides: Corsair Gorge"; says that "Captain Johnson of the *Columbus* evidently crossed the area many times with his echo-sounding apparatus running prior to the earthquake but never observed the valley"; and concludes: "It seems likely, therefore, that the valley either came into existence or was greatly extended at the time of the earthquake." The following year[21] the same interpretation is offered, with the further statement that "additional soundings made in the outer portion of the canyon in 1931 revealed hummocky topography characteristic of landslide accumulations."

The facts presented form an unsatisfactory basis for any conclusion that the gorge in question was opened by landslides precipitated by the earthquake of 1929. The position of the gorge, "about 500 miles from the epicentre of the earthquake and 340 miles from the nearest cable break," may perhaps raise doubt as to the competency of the distant shock to produce evacuation of a great gorge which had long remained filled despite previous earthquakes. That widely scattered soundings should not have happened to hit the gorge is admittedly no occasion for surprise. It is not shown that vessels crossing the Bank take numerous soundings in the immediate vicinity of the gorge. Vessels sounding merely to determine whether they have reached the shelf would not, from the nature of their soundings, discover the canyon. If the sounding is deep the navigating officer concludes that he has not reached the shelf, even if that deep sounding conceivably fell within the gorge. Once a shallow sounding proves the vessel is over the shelf, he does not need continuous further soundings to confirm a fact already known; nor does he normally make many soundings to work out the details of shelf topography. Only by an extraordinary combination of circumstances would a vessel in the ordinary routine of traversing a well-known ship lane make in succession a sounding on the shelf, another into the depths, and a third on the shelf, thus proving the existence of a canyon cut into the shelf surface. Continuous echo-sounding by skilled observers with good apparatus, made along traverses certainly crossing the gorge, should have revealed its presence. But the critical facts essential to establishing this important sequence are not presented in Shepard's discussion.

The sudden evacuation of vast quantities of debris from canyons miles in length and thousands of feet deep should seemingly result in vast accumulations at the canyon mouths. Material compact enough to fill such canyons to the brim, and thus prevent their discovery in the manner assumed by Shepard, could not be dissipated by distant flow over the ocean bed beyond the canyon mouths. Yet no evidence of such vast accumulations has been found. The reports of hummocky topography "in the outer portion of the canyon" (Corsair Gorge), and of "some hills on the floor of the valley such as would be expected to be left from a swirling mudflow" (Sagami Bay case, discussed below), must be accepted with reserve until the actual soundings are published with specification of the conditions under which they were taken. It is difficult to conceive of any ordinary submarine survey which would give data sufficiently abundant and sufficiently accurate to discriminate the minor details of landslide topography. Large accumulations of debris should readily be detected; and hills of moderate size could be recognized if soundings were adequate as to both quantity and quality; but it is not clear just what type of hills would be "left by a swirling mudflow." The elevation reported outside the submarine valley off the Columbia River, "which might perhaps be interpreted as a result of the sliding of material out of the valley and accumulating as a ridge beyond,"[22] belongs in the same category as other similar evidence referred to above, and cannot be given weight until the nature of the elevation is established.

The conception of great submarine canyons filled with debris, so as to escape detection until some recent earthquake caused their evacuation by landsliding or mudflowing, encounters formidable difficulties. There are several of these canyons on Georges Bank, none of which were discovered until recently. To fill these gigantic chasms far out on the edge of the shelf must have required a very long period of time, for in that location the filling process could have proceeded but slowly. We may perhaps reduce the time required if we place the filling in the glacial period and bring the melting ice sheet with its supply of debris to the vicinity of the Bank. But even under these conditions, filling under the sea must have required a considerable lapse of time. During this period there were presumably many earthquakes, possibly many more than have occurred along our northeast coast during historic time, because advance and retreat of the ice sheets apparently caused movements in the earth's crust which could either cause crustal displacements or precipitate them when initiated by other causes. We are required to believe, then, that during the long period of fill, and during the lapse of time since the fill, repeated earthquake shocks failed to keep the canyons cleared out; but that some few years ago a disturbance 500 miles away caused disgorgement of the fill on a truly titanic scale. Partial disgorgement of canyons incompletely filled will not suffice; for the whole case for landsliding in the Georges Bank instance rests on the assumption that the canyons were filled and hence not subject to discovery until after the 1929 earthquake. We must conclude that this assumption is of doubtful validity; and hence that the reported evidence of sliding is unconvincing.

Shepard has accepted in part but not wholly the evidence for depth changes in Sagami Bay during the Japanese earthquake of 1923, and believes these changes are partly due to landslides and mudflows.[23] In the second paper cited below he gives other supposed cases of depth changes during earthquakes and says: "These items of evidence appear to strengthen the landslide hypothesis as an explanation of submarine valleys." It should be observed that these reports of depth changes are based on comparisons of recent with older surveys. Such evidence is always suspect, because of the known inaccuracy of most early hydrographic surveys and because of the great difficulty (one might almost say the practical impossibility) of being sure that older and later soundings compared are sufficiently close to the same spot to serve as a reliable basis of comparison. Because the accuracy of surveys had previously been challenged in the case of the Japanese quake the writer while in Japan made some enquiry as to the materials and methods employed in determining the supposed depth changes. He reached the conclusion that the materials available afford no safe basis for depth comparisons, and that there is therefore insufficient evidence of the great depth changes reported. One may fully recognize that earthquakes can be accompanied by displacements of material on land or under the sea, and still doubt whether the Sagami Bay case, or other reported instances of marked deepening of submarine channels during earthquakes, lend valid support to the hypothesis that submarine valleys are either formed or re-excavated by landslides or mudflows.

As further support for the movement of landslides and mudflows in submarine canyons Shepard[24] reports the results of detailed comparative surveys made by him on the California coast. Surveys made in comparatively shallow water at the heads of canyons where these approach the coast showed appreciable changes. This was to be expected, since the extensive shifting of bottom debris in shallow water during constant grading and regrading of off-shore profiles has long been recognized. In the first of the two papers last cited Shepard says: "In each case the shallow water near the heads of the canyons was chosen as the best place for comparison." Measurements of depth changes are of course more easily and more accurately made near shore; and considerable changes in depth are most certain to occur there. But if the object was to determine the possible presence of landslides or mudflows, it would seem that no more unfortunate place could have been chosen.

That such was the object in view is indicated by the discussion of results obtained at Newport Pier and Redondo Pier. "The Newport Canyon was an ideal place for comparison since the Newport Pier has been built out to the edge of the canyon which cuts diagonally across its end." Again one must question the suitability of the location, as well as the form of the author's conclusion that "either mud-flows are carrying mud out of the canyon or [that] currents in the relatively shallow water perhaps combined with mud-flows are operating in this place." Since the author adds: "In view of the shallow water it is quite understandable that currents should exist here," and since currents in operation on every coast constantly scour and redeposit debris, there appears to be no reason to bring landslides or mudflows into the picture.

Currents moving past the end of a pier may be expected to exert an exceptionally marked scouring action there. Even in relatively calm weather the convergence of flow at such a point may well accomplish much deepening in a very short time. Shepard reports the occurrence of marked changes at the end of Redondo Pier where it "extends out into the head of a submarine canyon," fill taking place gradually at this point only to be followed by rapid deepening which changed depths of six feet to thirty-one feet. He concludes that the cause of removal of material from the several canyon heads from time to time, following its temporary deposition there, "judging from the Redondo Pier case must be of the nature of a submarine slip or flow." As in the case of the shallow water changes discussed above, there appears to be no reason to connect these normal near-shore changes with landslide or mudflow phenomena.

It thus appears that while much has been written about landslides and mudflows evacuating older canyons, there is little tangible evidence to support the conception that submarine slides and flows are of frequent occurrence, and none to connect them definitely with submarine canyons. It seems only reasonable to suppose that under favorable conditions slides may occur. But their bearing on the origin of submarine canyons for the present remains purely speculative.

Such speculation is valid when formulating a working hypothesis. And it is as a working hypothesis that one should examine the conception that ancient submarine canyons have been partially or wholly filled with debris, then later evacuated in whole or in part from time to time by landsliding, leaving the canyons revealed in their present form. Clearly the hypothesis suffers from obvious weaknesses when applied as a general explanation of submarine canyons. It rests upon a rather complicated series of unproved assumptions: that great quantities of loose sediment have been deposited in older canyons cut in more compact material composing the shelves on either side; that the sediments will be fluid enough to slide or flow at the slight disturbance of distant earthquakes of moderate intensity, yet compact enough to build up a fill sometimes thousands of feet in thickness; that the flowing and sliding can reach downward thousands of feet, and backward not only miles but sometimes tens or scores of miles, to clear canyons of great depth and length; and that such enormous fill can be evacuated without producing vast and hence easily recognized accumulations of debris in front of the canyon mouths. As already shown, there are no clearly demonstrated facts which appear to support the hypothesis. It remains, like other explanations of canyon origin, a working hypothesis offered in the effort to solve an exceptionally difficult problem. It is a hypothesis which apparently has been in part if not wholly abandoned by its author.

3. GROUNDWATER SAPPING

In order to explain the highly significant fact that many submarine canyons do not lie opposite the mouths of great rivers draining the continent, or at the ends of great channels cut across the continental shelf, Bryan[25] has suggested that the notches cut back in the face of the continental scarp may have grown by headward erosion through the agency of normal groundwater sapping when the region was above sea level. The minor side canyons tributary to the Grand Canyon of the Colorado are cited as examples of notches formed under such conditions.

While the headward extension of tributary canyons into a plateau is normally guided by shallow channels or other depressions on the upland surface, or by joints, faults, or other lines of weakness in the plateau rocks, it seems wholly possible for groundwater sapping to play an important rôle in the development of such canyons. Groundwater issuing as springs in valley heads is a normal accompaniment of the headward growth of valleys, and where the geological structure favors the undermining of higher beds, sapping is apt to occur. In any event the major point made by Bryan seems incontrovertible: notches cut many thousands of feet deep and a number of miles back into the face of a plateau scarp do not imply the former presence of large streams. Such notches are common features in the faces of mountain and plateau scarps the world over, and usually in circumstances which preclude the possibility of their erosion by streams of any considerable magnitude. They are the normal product of insequent and obsequent drainage, and of subsequent drainage guided by faults and other planes of weakness. The depth of the canyons depends on the height of the scarp, not on the size of the stream; and the length of the canyons depends on the length of time erosion has been in progress, not on the magnitude of the erosive agent.

It must be obvious, however, that the hypothesis of ordinary groundwater sapping above sea level does not touch the major difficulty presented by the deep notches in the face of the continental shelves. This difficulty is the great depth of their floors below present sea level. To find such notches in the walls of the Grand Canyon and in countless

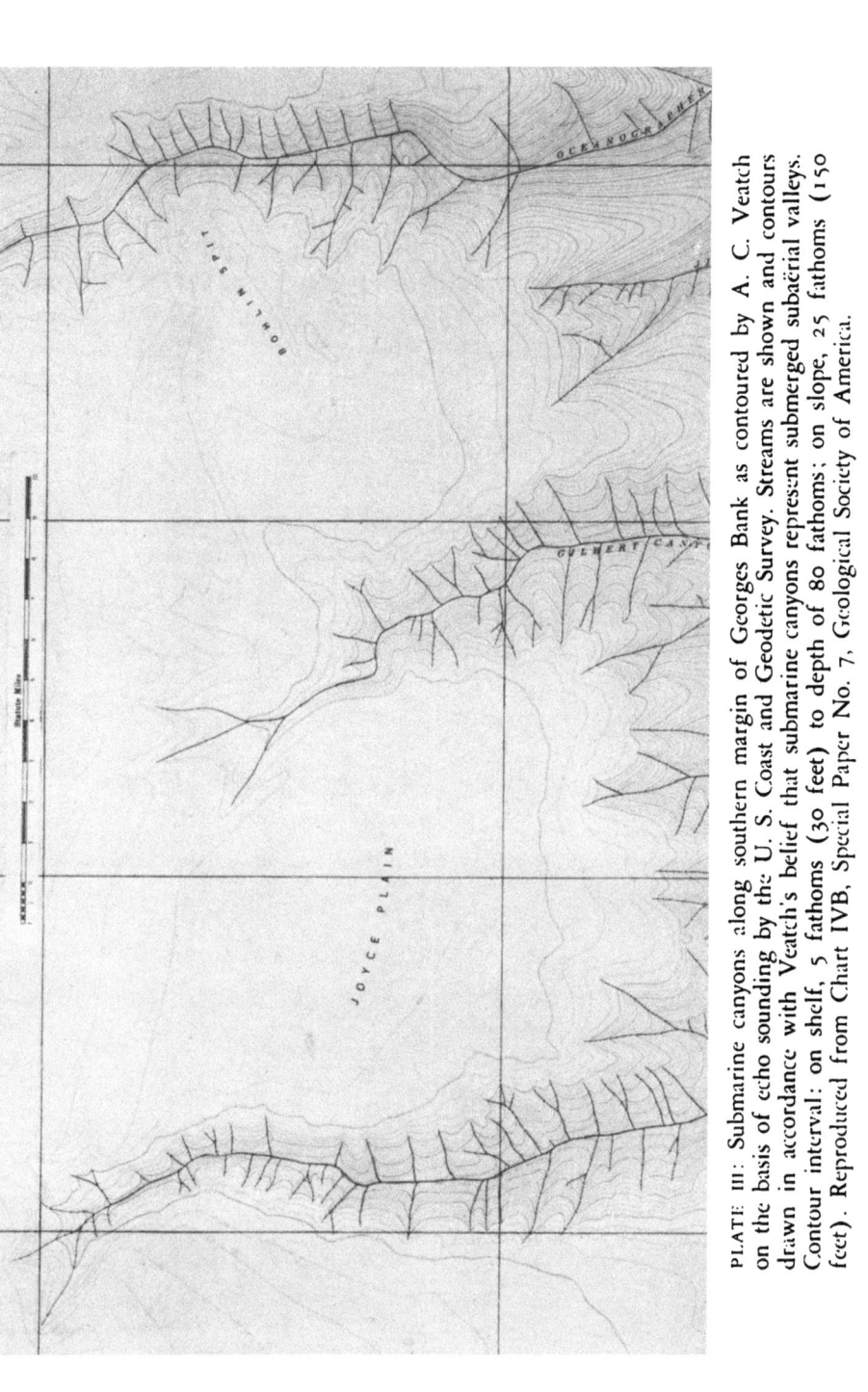

PLATE III: Submarine canyons along southern margin of Georges Bank as contoured by A. C. Veatch on the basis of echo sounding by the U. S. Coast and Geodetic Survey. Streams and contours drawn in accordance with Veatch's belief that submarine canyons represent submerged subaërial valleys. Contour interval: on shelf, 5 fathoms (30 feet) to depth of 80 fathoms; on slope, 25 fathoms (150 feet). Reproduced from Chart IVB, Special Paper No. 7, Geological Society of America.

other scarps on land creates no problem. To find them deep under the ocean creates a problem of immense difficulty for which no satisfactory solution has yet been found. The enormous vertical oscillations of land level or of sea level which must be postulated in any hypothesis involving a subaërial origin for the canyons, and the apparent absence of evidence confirming oscillations of such magnitude, have led many investigators to seek in submarine agencies a possible cause of the features in question.

Notes and References

1. As early as 1863, if not before, J. D. Dana attributed the submarine trench off the mouth of the Hudson River to subaërial erosion followed by submergence; but in his earlier work he seems to have discussed mainly if not wholly the shallow trench cut in the surface of the continental shelf. From the extensive literature ascribing this origin to deep shelf-margin canyons the following may be cited:

A. Lindenkohl, "Geology of the Sea-Bottom in the Approaches to New York Bay." *Amer. Jour. Sci.*, Vol. 29, pp. 475-480, 1885.

————"Notes on the Sub-marine Channel of the Hudson River and Other Evidences of Post-glacial Subsidence of the Middle Atlantic Coast Region." *Amer. Jour. Sci.*, Vol. 41, pp. 489-499, 1891.

Enrico Stassano, "La Foce del Congo." *Accad. dei Lincei Atti. Rendiconti*, Vol. 2, pp. 510-513, 1886.

A. Issel, "Sur l'existence de vallées submergées dans le golfe de Gênes." *Acad. Sci. Paris, Ct. Rend.*, Vol. 104, pp. 250-253, 1887.

J. D. Dana, "Long Island Sound in the Quaternary Era, with Observations on the Submarine Hudson River Channel." *Amer. Jour. Sci.*, Vol. 40, pp. 425-437, 1890.

J. W. Spencer, "The High Continental Elevation Preceding the Pleistocene Period." *Bull. Geol. Soc. Am.*, Vol. 1, pp. 65-70, 1890.

————"Terrestrial Submergence Southeast of the American Continent." *Bull. Geol. Soc. Am.*, Vol. 5, pp. 19-21, 1894.

————"Reconstruction of the Antillean Continent." *Bull. Geol. Soc. Am.*, Vol. 6, pp. 103-140, 1895.

————"Late Formations and Great Changes of Level in Jamaica." *Geol. Mag.*, N. S. Vol. 5, pp. 515-517, 1898.

————"Submarine Valleys off the American Coast and in the North Atlantic." *Bull. Geol. Soc. Am.*, Vol. 14, pp. 207-226, 1903.

————"The Submarine Great Canyon of the Hudson River." *Amer. Geol.*, Vol. 34, pp. 292-293, 1904.

————"The Submarine Great Canyon of the Hudson River." *Amer. Jour. Sci.*, Vol. 19, pp. 1-15, 1905; and in *Geog. Jour.*, Vol. 25, pp. 180-190, 1905.

————"The Submarine Valleys and Canyons off the American Coast." In Edward Hull's Monograph on the *Sub-oceanic Physiography of the North Atlantic Ocean*. (London) 41 pp., 1912. See pp. 21-30.

W. Upham, "The Fiords and Great Lake Basins of North America Considered as Evidence of Preglacial Continental Elevation and of Depression during the Glacial Period." *Bull. Geol. Soc. Am.*, Vol. 1, pp. 563-567, 1890.

————"Quaternary Changes of Levels." *Geol. Mag.*, Vol. 7, pp. 492-497, 1890.

————"Submarine Valleys on Continental Slopes." *Amer. Geol.*, Vol. 10, pp. 222-223, 1892.

W. Upham, "The Fishing Banks between Cape Cod and Newfoundland." *Amer. Jour. Sci.*, Vol. 47, pp. 123-129, 1894.

——"Fjords and Submerged Valleys of Europe." *Amer. Geol.*, Vol. 22, pp. 101-108, 1898.

Joseph Le Conte, "Tertiary and Post-tertiary Changes of the Atlantic and Pacific Coasts; with a Note on the Mutual Relations of Land-Elevation and Ice-Accumulation during the Quaternary Period." *Bull. Geol. Soc. Am.*, Vol. 2, pp. 323-330, 1891.

H. W. Fairbanks, "Oscillations of the Coast of California during the Pliocene and Pleistocene." *Amer. Geol.*, Vol. 20, pp. 213-245, 1897.

E. Hull, "The Submerged Platform of Western Europe." *Geol. Mag.*, Vol. 5, pp. 478-480, 1898.

——"Sub-oceanic Terraces and River Valleys of the Bay of Biscay." *Nature*, Vol. 57, p. 582, 1898.

——"Further Investigations Regarding the Submerged Terraces and River Valleys Bordering the British Isles." *Geol. Mag.*, Vol. 5, pp. 351-357, 1898.

——"On the Sub-oceanic Physical Features off the Coast of Western Europe, Including France, Spain and Portugal." *Geog. Jour.*, Vol. 13, pp. 285-289, 1899.

Fridtjof Nansen, "The Bathymetrical Features of the North Polar Seas with a Discussion of the Continental Shelves and Previous Oscillations of the Shore-Line." *The Norwegian North Polar Expedition 1893-1896, Scientific Results.* (London) Vol. 4, 232 pp., 1904. See pp. 95, 186-192, 231.

R. S. Holway, "Physiographically Unfinished Entrances to San Francisco Bay." *Univ. of Calif. Pub. in Geog.*, Vol. 1, pp. 81-126, 1914.

H. H. Hess, "Submerged River Valleys of the Bahamas." *Amer. Geophys. Union, Trans. 14th Ann. Meeting*, pp. 168-170, 1933.

John H. Maxson, "Structural Relationships of the Coast and Continental Margin of Northern California." (Abstract). *Bull. Geol. Soc. Am.*, Vol. 44, p. 152, 1933.

H. H. Hess and P. MacClintock, "Submerged Valleys on Continental Slopes and Changes of Sea Level." *Science*, Vol. 83, pp. 332-334, 1936.

F. P. Shepard, "Geological Mapping of the Ocean Bottom." *Science*, Vol. 82, pp. 614-615, 1935.

——"Continued Exploration of California Submarine Canyons." *Amer. Geophys. Union, Trans. 17th Ann. Meeting*, Part 1, pp. 221-223, 1936.

——"'Salt' Domes Related to Mississippi Submarine Trough." *Bull. Geol. Soc. Am.*, Vol. 48, pp. 1349-1361, 1937.

2. Reginald A. Daly, "Origin of Submarine 'Canyons.'" *Amer. Jour. Sci.*, Vol. 31, pp. 401-420, 1936.

3. F. P. Shepard, "The Underlying Causes of Submarine Canyons." *Nat. Acad. Sci., Proc.*, Vol. 22, pp. 496-502, 1936.

4. H. H. Hess and P. MacClintock, "Submerged Valleys on Continental Slopes and Changes of Sea Level." *Science*, Vol. 83, pp. 332-334, 1936.

5. F. P. Shepard, "The Origin of the Submarine Valleys along the Southern Edge of Georges Bank." *U. S. Coast and Geod. Surv., Assoc. Field Eng.*, Bull. 3, pp. 87-89, 1931.

6. F. P. Shepard, "Canyons South of Georges Bank." *U. S. Coast and Geod. Surv., Assoc. Field Eng.*, Bull. 6, pp. 39-42, 1932.

——"Landslide-Modifications of Submarine Valleys." *Amer. Geophys. Union, Trans. 13th Ann. Meeting*, pp. 226-230, 1932.

7. F. P. Shepard, "Canyons beneath the Seas." *Scientific Monthly*, Vol. 37, pp. 31-39, 1933.

——"Submarine Valleys." *Geog. Rev.*, Vol. 23, pp. 77-89, 1933.

8. F. P. Shepard, "American Submarine Canyons." *Scot. Geog. Mag.*, Vol. 50, pp. 212-218, 1934.

————"Canyons off the New England Coast." *Amer. Jour. Sci.*, Vol. 27, pp. 24-36, 1934.
9. F. P. Shepard, "Submarine Valleys." *Geog. Rev.*, Vol. 23, pp. 77-89, 1933.
10. Douglas Johnson, *The New England-Acadian Shoreline*. (New York) 608 pp., 1925. Evidence of limited marine abrasion is discussed throughout the work. See chapter summaries and Index.
11. F. P. Shepard, "Submarine Canyons of the American Coasts." *Zeit. für Geomorph.*, Vol. 9, pp. 99-105, 1935.
12. F. P. Shepard, "Geological Mapping of the Ocean Bottom." *Science*, Vol. 82, pp. 614-615, 1935.
13. F. P. Shepard, "Continued Exploration of California Submarine Canyons." *Amer. Geophys. Union, Trans. 17th Ann. Meeting*, Part 1, pp. 221-223, 1936.
14. F. P. Shepard, " 'Salt' Domes Related to Mississippi Submarine Trough." *Bull. Geol. Soc. Am.*, Vol. 48, pp. 1349-1361, 1937.
15. F. P. Shepard, "The Underlying Causes of Submarine Canyons." *Nat. Acad. Sci., Proc.*, Vol. 22, pp. 496-502, 1936.
16. John Milne, "Sub-oceanic Changes." *Geog. Jour.*, Vol. 10, pp. 259-284, 1897. See also R. D. Salisbury, "Mr. Forster on Earthquakes." *Amer. Geol.*, Vol. 3, pp. 182-188, 1889.
17. J. W. Gregory, "The Earthquake South of Newfoundland and Submarine Canyons." *Nature*, Vol. 124, p. 945, 1929.
18. F. P. Shepard, "The Origin of the Submarine Valleys along the Southern Edge of Georges Bank." *U. S. Coast and Geod. Surv., Assoc. Field Eng.*, Bull. 3, pp. 87-89, 1931.
————"Landslide-Modifications of Submarine Valleys." *Amer. Geophys. Union, Trans. 13th Ann. Meeting*, pp. 226-230, 1932.
————"Submarine Valleys." *Geog. Rev.*, Vol. 23, pp. 77-89, 1933.
————"Canyons beneath the Seas." *Scientific Monthly*, Vol. 37, pp. 31-39, 1933.
————"Depth Changes in Sagami Bay during the Great Japanese Earthquake." *Jour. Geol.*, Vol. 41, pp. 527-536, 1933.
Higgins (discussing paper by Hodgson and Doxsee), *East. Sect., Seis. Soc. Am., Proc. 1930 Meeting*, pp. 79-81, 1930.
D. S. McIntosh, "The Acadian-Newfoundland Earthquake." *Nova Scotian Inst. Sci. Trans.*, Vol. 17, Part 4, pp. 213-222, 1930.
E. M. Kindle, "Sea Bottom Samples from the Cabot Strait Earthquake Zone." *Bull. Geol. Soc. Am.*, Vol. 42, pp. 557-574, 1931.
19. F. P. Shepard, "The Origin of the Submarine Valleys along the Southern Edge of Georges Bank." *U. S. Coast and Geod. Surv., Assoc. Field Eng.*, Bull. 3, pp. 87-89, 1931.
20. F. P. Shepard, "Landslide-Modifications of Submarine Valleys." *Amer. Geophys. Union, Trans. 13th Ann. Meeting*, pp. 226-230, 1932.
21. F. P. Shepard, "Canyons beneath the Seas." *Scientific Monthly*, Vol. 37, pp. 31-39, 1933.
22. F. P. Shepard, "The Origin of the Submarine Valleys along the Southern Edge of Georges Bank." *U. S. Coast and Geod. Surv., Assoc. Field Eng.*, Bull. 3, pp. 87-89, 1931.
23. F. P. Shepard, "The Origin of the Submarine Valleys along the Southern Edge of Georges Bank." *U. S. Coast and Geod. Surv., Assoc. Field Eng.*, Bull. 3, pp. 87-89, 1931.
————"Landslide-Modifications of Submarine Valleys." *Amer. Geophys. Union, Trans. 13th Ann. Meeting*, pp. 226-230, 1932.
————"Depth Changes in Sagami Bay during the Great Japanese Earthquake." *Jour. Geol.*, Vol. 41, pp. 527-536, 1933.
24. F. P. Shepard, "Continued Exploration of California Submarine Canyons." *Amer. Geophys. Union, Trans. 17th Ann. Meeting*, Part 1, pp. 221-223, 1936.

F. P. Shepard, "Shifting Bottom in Submarine Canyon Heads," *Science*, Vol. 86, pp. 522-523, 1937.
25. Kirk Bryan, quoted by Henry C. Stetson, "Geology and Paleontology of the Georges Bank Canyons." *Bull. Geol. Soc. Am.*, Vol. 47, pp. 339-366, 1936. See p. 353. See also Henry C. Stetson, "Bed-Rock from the Continental Margin on Georges Bank." *Amer. Geophys. Union, Trans. 16th Ann. Meeting*, Part 1, pp. 226-228, 1935. See p. 226.

CHAPTER III

Hypotheses of Submarine Origin

1. SUBMARINE LANDSLIDES

It is quite within the limits of possibility that portions of the steep seaward slopes of continental shelves, especially where composed in whole or in part of unconsolidated sediments, should slump or slide into the deeps, or flow as masses of mud. The landslide scars might constitute perceptible notches in the shelf margins, but no known type of slide produces forms resembling the submarine canyons. As noted on earlier pages, Shepard has appealed to sliding and flowing as a means of evacuating ancient submarine valleys filled with more recent sediment, and has suggested that such a process might account for the deep outer portion of some canyons. Stetson and Smith[1] have shown experimentally that small and short "landslide valleys" with lunate heads are produced when water is lowered on the face of a miniature continental shelf; but they do not believe such a process can be accepted as a general explanation of submarine canyons. Daly[2] has considered the possibility that the canyons may be produced by turbidity currents (discussed later) accompanied by sliding in highly fluid sediments "triggered off" from a position of unstable equilibrium. But so far as the writer is aware no one would today depend upon landsliding or mudflowing alone to account for deep winding gorges many miles in length and cut in part at least in fairly well lithified sediments.

In a later chapter it will be shown that minor ravines or "furrows" on the steeper seaward face of a continental shelf may be the result of mudflowing during the deposition of sediment on the shelf face, or during consolidation of shelf sediment accompanied by the expulsion of connate waters in the form of springs. These minor ravines are not, however, to be confounded with the deeper canyons cut far back into the shelves. The origin of the minor forms is a distinct problem, for we have no proof as yet that they grow into major canyons, or have the same origin as the larger forms. One may therefore find support for the "furrowing" of a shelf face by mudflowing on a limited scale during shelf development, and yet find this agency wholly incompetent to excavate deep, steep-sided canyons many miles in length in lithified sediments. It is this latter hypothesis which apparently is not today seriously championed by any one, and which may therefore be dismissed without further discussion.

2. SUBMARINE CURRENTS

To explain submarine canyons appeal has frequently been made to sea-bottom currents, sometimes to erode gorges in previously deposited marine sediments, sometimes to form a gorge merely by preventing sedimentation along the course of a current while adjacent areas were being built up. Often the precise nature of the currents is not specified.[3] Such currents are supposed by some authors to move from the outer ocean toward the land, by others from the land seaward.

Various Types of Currents. When investigators have been somewhat more specific, the nature of the currents postulated have differed widely. Some have appealed to concentrated tidal currents;[4] some to seaward-moving bottom currents branching off from longshore littoral currents;[5] some to seaward-moving hydraulic currents flowing along the bottom when winds pile surface waters into coastal embayments.[6]

One variation of the submarine current hypothesis involves currents due to differences in density, flowing either landward or oceanward along the sea bottom. By some the density differences are ascribed to contrasts in salinity,[7] as where fresh water entering the sea sets up a seaward-moving surface current of lighter mixed fresh and salt water, above a landward-moving current of heavier salt water.*[8] Temperature differences and the presence of sediment in water have both (sometimes separately and sometimes in combination) been credited with causing currents capable of trenching the floors of lakes; and the possibility that currents of this type might cause submarine canyons of the type here discussed has been considered.[9]

All variations of the submarine current hypothesis encounter certain difficulties in common. Currents having a velocity too low to erode may be fast enough to prevent deposition; and where streams are building de-

---

* While the deep channel in the Skagerrak and Kattegat at the outlet of the Baltic Sea does not belong in the class of shelf-margin canyons, it is interesting to note that Ekman has attributed this channel to the scouring action of an inflowing salinity current when the land was higher and melting ice supplied larger volumes of outflowing water than now.

posits into lakes or quiet seas subaqueous channels of moderate magnitude sometimes result from the building up of material in the quieter water on either side of the entering currents. But it seems impossible to account for submarine canyons by a building up of the continental shelves for thousands of feet on either side of currents which at the end of the process must have been in many cases from one to two miles in depth and from one to several miles in breadth. The existence of such currents along the canyon axes is unknown. Channels of the type indicated are bordered by subaqueous ridges or levees which slope gradually downward as one passes away from the channel. Soundings have not revealed the presence of such levees along the canyon borders.

When one considers the great breadth of such a continental shelf as that bordering the northeast coast of North America, the relative smoothness of its surface, and the apparent absence of conditions favoring the development of localized currents of high velocity at its submerged outer border, one must doubt the existence there of currents sufficiently strong to erode the shelf sediments. Such doubts would exist even if the canyon walls revealed nothing but unconsolidated debris. They become multiplied manyfold as evidence accumulates that the canyons are cut, in part at least, in well-lithified material. As we shall show more fully on a later page, both Shepard and Stetson have discovered a high degree of lithification in some of the samples dredged from submarine canyon walls. When discussing the origin of such material Stetson wrote in 1935: "It is almost impossible to conceive of hydrographic conditions which could produce submarine currents powerful enough to cut such gashes."[10]

The doubts based on deductive reasoning are intensified by the fact that currents of the types mentioned and of the strength required have not been discovered along shelf margins or in the canyons. During the summer of 1936 measurements were made in the bottoms of three of the remarkable canyons of Georges Bank, in order to determine what velocities might there exist in "normal currents due to differences in densities caused by temperature and salinity," and in "those which have a tidal origin." Sediment density currents which might result from peculiar physical conditions no longer existing could not, of course, be studied. Stetson's report[11] of the results secured confirms the view of most students as to the limited competency of normal submarine currents, which in this case were predominantly tidal. If movements due to temperature or salinity differences were present, they must have been so weak as to be negligible. Stetson writes:

The velocities recorded in the canyons are the normal ones which occur on the shelf under present-day tidal conditions. These currents, therefore, are merely flowing in preexisting cuts and are not the cause of them. It does not seem probable that they could be particularly competent agents of erosion when attacking the more indurated material which we know makes up part of the valley-walls. If this be the case in a region of strong tidal currents such as Georges Bank, they would be even less effective farther south where smaller velocities might be expected. . . . . On the basis of the above evidence, we are justified in eliminating from further consideration present-day bottom-currents as forces which are capable of forming the east-coast canyons. They may have helped to preserve them, but that is the full extent of the part which they have played. Nor does it seem likely that the type of currents which we have been discussing behaved in the past very differently from the way that they are doing today.

Many observers have reported the presence of very fine silt on the floors of submarine canyons. This fact has generally been held to justify the conclusion that no currents operating at the shelf margin under present hydrographic conditions could have been responsible for carving the deep trenches. Davis has suggested that such fine sediment as occurs may have been deposited under present conditions *since the last violent storm*.[12] One must hesitate to adopt this suggestion when the canyons are located far out on the deeper edge of a broad continental shelf, and when silt is found in considerable thickness 3,000 feet or more below sea level.[13] Davis's interpretation clearly is not applicable when, as in the east-coast canyons, lower portions of the silt contain an Arctic fauna with species not now living in the surface silt.[14]

That such deposition could occur at any time seemingly excludes effective erosive action by normal currents due to differences in temperature, salinity, or tidal forces. Outflowing hydraulic currents during heavy on-shore winds might operate locally where waters are banked up at the head of a coastal embayment; but it seems impossible to invoke them in case of the many canyons found at the outer edge of broad and relatively smooth continental shelves, or off straight coasts.

It thus appears that if we appeal to submarine currents as a general cause of shelf-margin canyons, we are thrown back on hypothetical currents due to the superior density of water loaded with sediment. Since no effective bottom currents of this type are known to exist off the mouths of silt-laden rivers or in other places supposedly favorable to their development, we must further assume, as Daly[15] has done, that in some past time, such as the glacial period, hydrographic conditions were far more favorable than now to the development of such currents. This opens up a problem to which very careful consideration must be given.

Turbidity Currents. It has long been recognized that turbid water is, by virtue of the sediment held in suspension, of greater density than the same water when clear; just as saline water is, by virtue of the salt held in solution, of greater density than the same water when pure. Currents resulting from these differences in density belong in the general group of density currents, but need to be distinguished by special names. Those due to salinity have already been called *salinity currents.* By analogy those due to turbidity will here be called *turbidity currents.* Because turbidity results from material held in suspension, the name "suspension currents" was recently suggested for currents of this origin.[16] The corresponding name for currents due to material held in solution would be "solution currents," a name obviously objectionable. As currents of either type entering the sea will flow down the continental slope until a layer of greater density is reached, and will then spread horizontally, suspended between the lighter water above and the heavier water below, there is possibility of confusion in using the term "suspension" as a name for currents of either type. "Turbidity currents" is clear as to meaning, and analagous with "salinity currents" already in use.

Special prominence has recently been given to turbidity currents through the important paper by Daly[17] on the "Origin of Submarine 'Canyons'" previously cited. Under the subtitle "A New Conception of Origin" Daly elaborates the idea that waves beating upon the outer part of the continental shelf during the lowered sea level of Glacial time charged the sea water with sediments to such an extent that its density was increased, the heavier water then flowing down the shelf slope and over its steep border in currents which, aided by hydraulic currents generated by the piling up of water by on-shore winds, carved the canyons. That there are serious objections to this conception of canyon origin Daly fully recognizes, and he supports the conception merely as a working hypothesis seemingly less drastic than the drowned valley hypothesis which necessitates apparently incredible recent changes in the level of land or sea.

In fairness to earlier investigators it should be pointed out that there is nothing new in the conception that currents charged with sediment may become heavy enough to descend into less dense clear water, in the conception that such currents may produce subaqueous trenches of great magnitude, or in the conception that currents of this type, aided by ordinary hydraulic currents, carved submarine canyons during the lowered sea level of Glacial time. Even the striking analogy with flowing mercury, employed by Daly to illustrate the process, appears repeatedly in the foreign literature. Let us step back more than half a century and review the development of these conceptions.

In the years 1883 and 1885 the engineer J. Hörnlimann, while preparing hydrographic maps of Lake Constance and Lake Geneva for the Swiss Hydrographic Bureau, discovered that where the Rhine enters the first and the Rhone enters the second of these lakes, a remarkable subaqueous trench exists in the deposits off the mouth of each stream. The sublacustrine trench of the Rhine was described as being 4 kilometers in length (later found to be much longer), 600 meters in breadth at the widest; and up to 70 meters in depth below the lake floor and 140 meters or more below lake level; that of the Rhone as more than 6 kilometers long, 500 to 800 meters wide, with a depth ranging up to 50 or 60 meters below the lake floor and 230 meters below lake level; the maximum depths of the trenches below the adjacent lake floor being closer to the river mouths than their maximum depths below lake level.

Immediately these subaqueous trenches attracted the attention of others, and speculation as to their origin began. It had long been a matter of observation that the muddy waters of the Rhine and Rhone disappear suddenly and with great commotion under the clearer waters of the lakes into which they empty. The common explanation had been that the sudden sinking of the river water below the lake water was a result of temperature differences, the glacier-fed streams being colder and hence heavier than the waters of the lower-lying lakes. It was but a short step to connect these phenomena genetically.

In 1884 that step was taken by a distinguished engineer of Dutch birth, Adolf von Salis-Soglio.[18] When this authority, who was chief engineer and inspector of works for the Swiss Confederation, learned of Hörnlimann's 1883 discovery of the trench at the mouth of the Rhine in Lake Constance, he concluded that the heavier cold river water sank to the floor of the lake, and continued as a bottom current which eroded the observed trench in a subaqueous cone or delta previously deposited by the stream. As von Salis accepted the prevailing view that temperature differences caused the river water to sink in the lake, he was not impelled to seek an explanation in turbidity currents.

A year later Forel,[19] who apparently had not yet seen the article by von Salis, published his first paper on this subject, one of two destined to become classic and to be quoted in scientific reports, reference works, and even in college textbooks from that day to this. Forel showed that the trenches in the sublacustrine deltas of the Rhine and Rhone were really depressions between dykes or levees which rose on either side. He concluded that the trenches resulted in part from erosion by a bottom current, and in part from deposition of material to form the dykes in the quieter water on either side of this current. "This deep current results from the greater density of the river waters, which are heavier than the lake waters (1) because of their [lower] temperature, and (2) because of the load of sediment which makes them milky." Thus the principle of the turbidity-density current was brought into the picture by Forel, but apparently as playing a secondary rôle. Emphasis was placed chiefly upon temperature differences, and it was argued that throughout the summer the river water was colder than the surface waters of the lake, while in springtime they were colder than even the deepest lake waters. "The glacial alluvium renders still heavier these river waters."

Two years later Forel[20] returned to discussion of the curious trenches found in the sublacustrine deltas of the Rhine and Rhone, and on the basis of further study materially modified his earlier opinion of their origin. Still holding that temperature differences alone will cause the river water to descend some distance below the lake surface during part of the year, he concluded that further increase of density due to mineral matter held in solution and to sediment held in suspension is necessary to enable the river current to reach the greater depths in which distal parts of the trenches are found. Thus he held that the effective currents responsible for the trenches were due to a favorable combination of turbidity, salinity, and temperature differences. More important still, he recognized that the borders of such a current, where the flowing river water is in contact with the stagnant lake waters, must be areas of deposition; and that up-building of the lateral dykes would alone suffice to give the intervening trench. "We have no necessity to invoke erosive activity, as I believed in my first studies." He recognized that some erosion is possible, perhaps even probable in the upper (nearer the river mouth?) parts of the trenches; but considered that erosion is unlikely in those parts of the trench far out from the river mouth and from 100 to 200 meters below lake level. "There we have, without doubt, only deposition."

In a still later work Forel[21] discussed at length the question as to whether the Rhine and Rhone trenches represent (a) the result of erosion, (b) the result of deposition on either side, or (c) original topographic depressions not completely obscured by deposition. After disposing of the first and third possibilities, he concluded that the trenches result from non-deposition along the axis of the current, while deposition active in the stagnant water on either side builds up lateral dykes. To illustrate the flow of the bottom current of river water, found to be heavier than the lake water most of the year when account was taken of "the temperature of the river water and its load of dissolved and suspended alluvion," Forel in this same work (*Le Léman*) employed the curious analogy of a river of mercury or of sulphuric acid ("comme le ferait un fleuve de mercure ou d'acide sulfurique qui descendrait dans le lac").

Following Forel many writers discussed the sublacustrine trenches of the Rhine and Rhone, usually accepting his explanation of their origin, but sometimes modifying or adding to it. Wey, Eberhard, Delebecque, Heim, Kleinschmidt, Collet, and Schmidle made contributions which will be mentioned when details of the turbidity hypothesis are discussed. In all three of the papers by Delebecque, cited on later pages, that author repeats Forel's curious analogy with the subaqueous flow of a river of mercury.

The origin of submarine canyons bordering the continents had been discussed for at least a quarter-century before Forel wrote, and was attracting further attention at the time of his writing. It is hardly to be supposed that this acute observer could fail to consider the possibility that the combined turbidity-salinity-temperature currents postulated by him to explain the sublacustrine trenches at the mouths of the Rhine and Rhone might likewise explain the submarine canyons off the mouth of the Congo, in the Bay of Biscay, and elsewhere. We are not surprised, therefore, to find that in his 1887 paper, cited above, Forel specifically refers to the "analogous trenches" known to occur beneath the ocean opposite the present or ancient mouths of rivers, but on a grander scale than those found in the Swiss lakes. Admitting that he knew too little of the physical conditions which might explain the great submarine canyons to justify his discussing the theory of their origin, he added: "Their theory is evidently very different from that of the sublacustrine trenches which occupy our attention." In one of his later works Forel[22] returned to the discussion of submarine canyons, and reiterated his opinion that the theory competent to explain these great suboceanic gorges must be quite different from that proposed by him to explain the sublacustrine trenches of the Rhine and Rhone.

If Forel was the first to consider the question as to whether currents due to the combined action of turbidity, salinity, and temperature differences were competent to explain submarine canyons, he was by no means the only one. In 1891 Angelo Heilprin[23] rejected the Forel theory as a satisfactory explanation of the Hudson submarine trench. A year later Linhardt[24] discussed at length the problem of submarine canyons, referred to the density-current theory offered in explanation of the sublacustrine trench of the Rhone, and concluded that even if we assumed the existence of similar currents in the ocean, these could not alone account for erosion of the great suboceanic gorges. In his textbook, "Introduction to Geology," Scott[25] quotes as exceptional the special conditions found where the Rhine and Rhone enter Lake Constance and Lake Geneva, and implies that they cannot be invoked to account for submarine trenches like that of the Hudson.

Davis was apparently the first to correlate the work of turbidity currents with Glacial changes of sea level. In 1933 this author mentioned submarine canyons off the California coast in a discussion of Glacial changes of sea level in that region, and made the following statement: "These 'mock valleys,' as I proposed to call them, should therefore be ascribed to the scouring action of marine agencies during the recent rise and advance of the sea when the present cliffs were cut back. This topic will be treated elsewhere in a special article."[26] In such an article published the same year Davis wrote: "The sigmoid pattern of the mock-valley may thus represent the course of a temporary submarine current driven seaward while the surface-waters are driven landward by on-shore storms. It is possible that the shore-waters would at such times gather a load of sediments which would increase their specific gravity and thus aid them in descending into the cold waters of the depths."[27] A year later Davis described the submarine canyons or mock valleys of this region more fully, showed that the configuration of the shore in the vicinity of the Monterey submarine canyon was especially favorable to the development of a seaward-moving hydraulic current during on-shore storms, and added: "It is possible that the shore waters, made turbid at such times by wave action, would thus gain an increased specific gravity that would facilitate their descent into the colder water of the depths. . . . . It is inconceivable that such a feature [Monterey submarine canyon] could have been kept open without the aid of a storm-driven, out-sweeping current during the last Glacial lowering and rising of sea level."[28]

Three years after publication of Davis's first paper on the subject, Daly[29] elaborated much more fully the density-current hypothesis. Like Davis, Daly attributed the increased density to turbidity resulting from wave action during storms, recognized that wind-generated hydraulic currents would coöperate with those due to differences in density, and correlated the work of such currents with the changed sea level of Glacial time. But whereas Davis seems to have assigned a subordinate rôle to turbid waters descending toward the deeps because of increased density, stressing rather the wind-generated hydraulic currents the descent of which would be "facilitated" by density differences, Daly gave a subordinate place to the hydraulic currents, stating that they would "accelerate" currents arising from greater density of turbid water. Daly further assumed that during the lowered sea level of Glacial time "wind waves and tidal waves were beating on the mud and sand of the continental shelves—a condition utterly unlike that now ruling," and believed such condition enormously facilitated the development of turbidity currents. As Daly's paper constitutes the strongest case yet made in favor of the development of submarine canyons by turbidity currents, we shall consider it more fully on later pages.

Shortly after publication of Daly's paper, Shepard[30] contributed an extended criticism of the turbidity-current hypothesis of submarine canyon origin. Paul A. Smith, Jr.,[31] stated that there are serious objections to the hypothesis "from an engineering as well as a physical viewpoint," and showed[32] that in the vicinity of Bogoslof Island, where there are no extensive coastal shelves such as figure prominently in Daly's version of the hypothesis, there are some remarkable submarine canyons; whereas farther north, where a comparatively smooth submerged coastal slope

Submarine Currents

offered ideal conditions for development of turbidity currents during the lowered sea level of Glacial time, submarine canyons are either lacking or poorly developed. On the other hand, Kuenen[33] reported the results of experimental studies which in his opinion are favorable to the conception that turbidity currents may be responsible for the development of submarine canyons. Experimental and other studies by H. C. Stetson and J. Fred Smith[34] lead them to conclude that turbidity currents in a stable ocean will flow down the continental slope until they reach a layer of the ocean having greater density than their own, when they will spread out horizontally over the top of the heavier layer. "Therefore, regardless of the ability of currents of this type to erode, they may be important agents for the distribution of sediment over the continental slope and nearby ocean floor."

Such are some of the milestones in the development, during the last half-century, of the hypothesis of turbidity currents as a factor in the formation of sublacustrine trenches and submarine canyons. We shall next examine the strong and weak points of this hypothesis as applied to explanation of the latter forms.

Strength and Weakness of the Turbidity Hypothesis. The points favorable to the turbidity-current hypothesis of submarine canyon origin have been most ably marshaled by Daly.[35] With characteristic skill he pictures the conditions believed by him to exist on continental shelves during Glacial time, and illustrates the consequences which in his opinion must have resulted:

Each of the four sets of ice-caps grew slowly and melted slowly, each process taking at least 25,000 years. Hence for more than 200,000 years, out of the million years or so that have elapsed since the Ice Age dawned, wind waves and tidal waves were beating on the mud and sand of the continental shelves—a condition utterly unlike that now ruling. These more or less mobile sediments had been built into embankments with widths measuring scores of kilometers and with depths averaging at least tens of meters. The volume of fine sediments was therefore enormous and sufficient to keep the tidal currents and storm waves of the lowered ocean well charged with solid particles for a large fraction of the 200,000 years. The waves were especially muddy because the depth of water on the outer, still submerged parts of the shelves was small. Then, too, the average storminess of the world was doubtless more pronounced during the Glacial epochs than at present. Storms no more intense than those now affecting the shelves must have made the water overlying the continental slope (the fall-off of each shelf) much richer in suspended sediment than the water of similar location in pre-Glacial, inter-Glacial, or post-Glacial times. The tidal currents and gales of the twentieth century disturb the bottom of the North Sea so powerfully that sand is thrown up from depths of 40 to 50 meters to the decks of laboring ships. So long as sediment was "suspended" in the water on the Pleistocene shelves, that water was effectively denser than the clean water farther out to sea or the water below the zone of rapid stirring. There must have been a tendency for the weighted water to dive under the cleaner water, to slide along the gently inclined bottom of the shelf, and to flow still faster down the steeper continental slope. Since the solid particles kept settling out, the horizontal distance through which any such density current operated was limited. It is therefore important to remember throughout the discussion of the general hypothesis, that the belt of strong agitation by waves was, at the times of lowered sea level, much nearer to the continental slope than now. In principle the imagined bottom current would be similar to the flow of ink or muddy water placed at the appropriate point in a tilted, partly filled glass of clear water. Each of those denser fluids slides down along the inclined "floor" of containing glass.

If a thick, uniform sheet of mercury were kept continuously pouring out from the whole shore of a continent, it is easy to see what would happen in a general way. The mercury would quickly seek and flow down any slight, initial, transverse depression in the continental shelf and erode the depression deeper. With this progressive deepening, more and more mercury would be drawn into the new trough from both sides, thus increasing the thickness, velocity, and eroding power of the stream of mercury that occupies the trough. To him "that hath shall be given." Reaching the much steeper continental slope, the mercury flood, so concentrated, would slide down yet faster and dig a deeper trough in the soft sediments beneath the slope. Continued long enough, a submarine "canyon" would there be produced. This fanciful analogy may help to clarify the hypothesis now to be elaborated.

Daly then presents evidence and arguments to show that under the conditions pictured, large quantities of sediment could be taken into suspension during storms, or contributed by settling from overlying muddy river water; that much of this sediment would settle very slowly even in sea water; that the resulting increase in effective density would be sufficient to develop submarine currents of considerable velocity; that the drag of such currents might initiate the seaward sliding of mobile, water-soaked mud on the sea floor, with consequent mixing of the agitated mud and original currents, and a resulting further increase in the drag and cutting power of the latter; that the turbidity currents would be further accelerated by seaward-moving hydraulic currents when strong on-shore winds piled sea water upon the coastal flats; and that the combined force of these currents might be sufficient to excavate submarine canyons in poorly lithified sediments of the continental shelves during Glacial time, and possibly to some extent also in pre-Glacial time.

No limited quotations from, or summaries of, Daly's article can do full justice to the plausibility of the turbidity-current hypothesis of canyon origin as presented by that author. The reader must study the original text to appreciate fully the best case that has been made (and probably that can be made) in favor of the competence of turbidity currents to erode suboceanic gorges of vast magnitude. Let us now consider some of the weaknesses of this hypothesis of canyon origin.

Sublacustrine Trenches of the Rhine and Rhone Rivers. First we must note that there are in Daly's presentation of evidence and arguments important errors and omissions which make his case in favor of turbidity currents appear much stronger than it really is. This is notably true in his discussion of the sublacustrine trenches of the Rhine and Rhone river deltas in Lakes Constance and Geneva, and the evidence they offer as to the competence of turbidity currents to cut subaqueous gorges of great magnitude. After discussing the trench of the Rhone, Daly writes: "In his first paper on the subject Forel *supported the conclusion* of de Salis, that the trench was *cut* (creusé) by the bottom current, *flowing because of the load of silt and consequent excess of density.*"* In this sentence the weight of the two authorities quoted would appear to support the turbidity-current hypothesis of trench erosion. But so far as the writer can discover: (a) von Salis did not believe the bottom current was flowing because of the load of silt and consequent excess of density; (b) Forel himself did not support the view that the bottom current resulted solely from the load of silt and consequent excess of density; and (c) Forel never accepted von Salis's view that the trench of the Rhine (the

---

* Italics by the present writer.

only trench discussed by von Salis), or that of the Rhone, was due solely to erosion.

(a) The reference to de Salis, apparently based on the French of Forel, obviously concerns the paper by von Salis which was cited in note 18 above. Daly does not refer directly to this paper, but cites Forel's first (1885) paper on the Rhine and Rhone trenches. In the first paper, however, there is no reference to von Salis. At that time Forel apparently had not seen the article by von Salis, and gave an independent explanation of the trenches. Two years later (1887) Forel published his second and principal essay, in which the work of von Salis is quoted, but not in the sense given by Daly. Because of an apparent confusion of references in Daly's paper the reader cannot be sure what were the particular statements Daly intended to cite, nor whether he saw the original form of Forel's principal paper with its reference to von Salis.

In any case the views of both von Salis and Forel as set forth in the literature cited in the present volume are quite clear. In his 1887 paper Forel quotes von Salis, in free translation, as concluding that the trench of the Rhine was cut by a bottom current formed when the river water sinks to the bottom of the lake, this sinking being due to *the difference in temperature of the two waters* ("et l'on peut attribuer cette chute à la différence de température des deux eaux").[36] Forel correctly quoted von Salis who, as already noted on an earlier page, made no reference to any excess of density due to the load of silt carried by the river water, but accepted the prevailing view that the waters of the Rhine and Rhone sank because they were colder than the lake waters. In the words of von Salis: "Es dürfte daraus mit Recht geschlossen worden sein, dass das Flusswasser an der Mündung versinke, was man sich bekanntlich aus der Temperaturverschiedenheit erklärt hat."[37]

While later studies have demonstrated that temperature differences are not, as von Salis believed, the sole factor in causing the Rhine and Rhone waters to sink in their respective lakes, they constitute one of three factors all of which combine during part of the year to make this sinking possible. A correct appreciation of this fact becomes of vital importance when one tries to establish an analogy between the known bottom currents in Lakes Constance and Geneva, and hypothetical bottom currents in the ocean where temperature differences constitute one of two factors, out of the three involved in the lakes, which tend to *prevent* analogous bottom currents.

(b) Forel did not, as one might infer from reading Daly's paper, support the view that the bottom currents believed to be responsible for the Rhine and Rhone trenches resulted solely from the load of silt carried by these streams and the consequent excess density of their waters. He held, rather, that the bottom currents were due to excess density resulting from a combination of three factors, one of which (material held in solution) varied but little throughout the year, while the other two (temperature differences and material carried in suspension) were negligible or non-active at some seasons, highly important in others. Thus he found that from late fall to early spring the amount of material carried in suspension by the Rhone was negligible, being even less than that carried in solution; but that from late spring to early autumn the quantity of material carried in suspension was enormous, greatly overbalancing all other factors combined, but generally coöperating with them in determining the excess density of the river water.

In his discussion of Forel's hypothesis Daly makes no mention of the fact that this investigator attributed considerable importance to the effects of differences in temperature between the river and lake waters. Instead, he says that "if the hypothesis [of turbidity currents] is well founded, the water of rivers that are building deltas *into fresh-water lakes of nearly the same temperature** should plunge under the water of the respective lakes" when these rivers carry much sediment; cites the entrance of the muddy Rhone into Lake Geneva as an example; and presents only that part of Forel's explanation which has to do with turbidity currents. Yet Forel gave much time and labor to detailed studies of temperature conditions in the Rhone River and in Lake Geneva; and his major paper (1887) on the origin of the trenches devotes much space to demonstrating the existence of important contrasts in temperature between the river water and the lake water; contrasts which, without intervention of other factors, in winter would cause the river water to spread out over the lake water, but during summer would cause the river water to sink in the lake to variable depths up to 40 meters, thus at that season coöperating with the effects of material carried in solution and in suspension to give a bottom current reaching to much greater depths.

In his earliest paper[38] Forel gave temperature contrasts first place as a cause for sinking of the river water; and in his later works[39] he con-

---
* Italics by present writer.

tinued to assign an important rôle to temperature differences, even when recognizing that at certain seasons these alone may temporarily be unfavorable to the generation of bottom currents. Seldom does he mention such currents without including reference to the element of temperature. "Les eaux froides et chargées d'alluvion" is typical of his expressions, even in his later works. Hence to omit all references to these temperature differences when citing Forel's explanation is to give a very incomplete idea of that writer's views, and to make him assign greater potency to turbidity currents than he actually did.

Not only Forel but most of those who have since discussed the origin of the Rhine and Rhone trenches have emphasized the importance of temperature differences and other factors in helping to account for the sinking of the river waters below the lake waters to give rise to a bottom current. Forel[40] studied not only temperature differences but also differences in the quantity of mineral matter found dissolved in river and lake, in his effort to explain the sinking of the Rhone waters. He found that throughout the year the Rhone water had a content of dissolved salts appreciably greater than that of Lake Geneva, as a result of which the river water from that cause alone must have a density slightly greater than that of the lake water. Eberhard Graf von Zeppelin[41] accepted the view that the combination of lower temperature and greater turbidity caused the Rhine and Rhone river waters to sink "under the warmer and lighter water of the lakes." Delebecque[42] is an exception to the general rule in that he usually mentions turbidity alone when discussing the origin of the Rhine and Rhone bottom currents, although he affirms his acceptance of Forel's ideas, and points out that river water rendered more dense by material in solution rather than by suspended sediment would behave in the same manner. Heim[43] repeatedly places first emphasis upon temperature differences, with differences in quantity of material in solution and of material in suspension following in the order named. Heim further shows that the turbid waters of certain streams are, especially in late summer or early autumn, lighter than the waters of those lakes into which they empty; and that under these circumstances the turbid water spreads over the lake surface as a floating muddy film, with the result that no sublacustrine trench is formed or, if temporarily begun, is soon filled and thus obliterated. Kleinschmidt,[44] discussing the Rhine trench, appears to consider as most important the "relatively low temperature" of the Rhine water, and as of secondary importance its content of material. He believes that in summer the river water sinks only part way toward the bottom, and that then deposition occurs everywhere in front of the river's mouth; but that in the short winter season, when alone, in his opinion, the river water reaches into the depths, the fine deposits are removed along the bottom and sides of the current to give a trench which thus owes its existence in part to up-building of lateral areas and in part to intermittent erosion of the channel bed. More recently Schmidle[45] repeatedly mentions temperature conditions alone when accounting for the sinking of the Rhine water in Lake Constance, but limits this sinking to 100 meters. He explains the deeper part of the trench as an unfilled tectonic depression, an explanation early advanced by Du Parc[46] and others but rejected by most students of the Rhine and Rhone trenches.

The foregoing citations from the literature are sufficient to show that there has never been complete agreement as to when the bottom currents in Lakes Constance and Geneva are active, to what depths they descend, or what factor is dominant at different seasons in causing the Rhine and Rhone river waters to flow as currents on the bottoms of the lakes. There is not even substantial agreement that turbidity is the dominant factor during the season when the Rhine and Rhone are most heavily charged with sediment. The fact, observed by Heim and many others and further discussed below, that many turbid streams enter lakes without sinking to the bottom and without producing subaqueous trenches is evidence of the strongest kind that factors other than turbidity may be controlling as to whether river or lake will have the superior density. Under these circumstances it is manifestly inadmissible to ignore those factors other than turbidity which were invoked by Forel and his followers in their efforts to explain the sublacustrine trenches of the Rhine and Rhone deltas. Especially is this true when the origin of these trenches is cited in support of a hypothesis of origin for submarine canyons which must involve these same factors, but under conditions in which certain of the factors will tend to prevent, instead of to produce, the required bottom currents.

(c) While von Salis did believe that the Rhine trench is due solely to erosion by a bottom current, that view was never accepted by Forel. In his first paper on the subject, apparently written before he had seen von Salis's article, Forel attributed the origin of the trenches partly to erosion along the axis of the current and partly to deposition of lateral dykes in the stagnant water on either side. After reading von Salis's article, in which the channel was attributed wholly to erosion, Forel published his second paper, showing that no erosion was required to account for the observed facts, although the possibility of some erosion in favorable localities is recognized. Still later, in his great work on *Le Léman*, he seems even more firmly committed to the view that the sublacustrine trenches are phenomena of deposition and not of erosion.

Daly recognizes that Forel's latest view was unfavorable to the erosion hypothesis; but he does not give the evidence which led that investigator to abandon even that measure of erosion he was first willing to admit. This evidence is highly pertinent to any discussion of the turbidity-current hypothesis of submarine canyons which seeks support in the phenomena of the Rhine and Rhone trenches. For, as Daly is careful to point out, there are serious objections to the conception that differential deposition of silt may account for submarine canyons. If Forel was correct in his belief that the sublacustrine trenches of the Rhine and Rhone are the product of such differential deposition, arguments in favor of the turbidity-current hypothesis of submarine canyons, based on analogy with these trenches, lose their force.

Forel states most fully in his *Le Léman* the considerations which impelled him to abandon his earlier view that the trenches might be in part a phenomenon of erosion. He shows that the bottom current of river water, flowing between walls of stagnant lake water, must set up eddies on either side in which the river water will mix with the lake water, lose its velocity, and deposit its burden of silt. There must thus result two parallel dykes or levees between which the bottom current of river water will be contained. The lateral dykes are shown by soundings to have the form of typical levees of deposition, sloping gently downward on their outer sides, but more steeply on their inner sides, facing the current. The dykes stand in relief on the delta surface. Furthermore, in the case of the Rhone trench, the bottom of the so-called ravine has about the same elevation as the general surface of the delta beyond the limits of the dykes. As Delebecque later expressively phrased it, "If one imagines the two dykes which limit the ravine to be removed, there would remain only a scarcely perceptible furrow on the surface [of the delta]." In the case of the Rhine trench the bottom of the depression is described by Forel as sensibly below the surface of the delta.

It thus appears that both theoretical considerations and observed facts support Forel's conclusion that the sublacustrine trenches of the Rhine and Rhone are depositional phenomena and not the product of erosion. The levee-like form of the dykes and the facts that they stand in relief above the general surface of the deltas and that the bottom of the channel in one case scarcely reaches below that surface testify strongly to the correctness of Forel's interpretation. That the bottom of one trench should extend sensibly below the adjacent delta surface is, as Forel recognized, wholly compatible with the idea that the trench is entirely the product of deferred deposition along the axis of the sublacustrine river current.

Forel's conclusion that the sublacustrine trenches of the Rhine and Rhone are purely depositional phenomena apparently was anticipated by Wey,[47] who in 1887 wrote: "While the opinion prevails among various scientists that it [the Rhine trench] has been produced through scouring ("auswaschung") I am of the opinion that it was formed through the lateral deposition of sediments, and remained open because deposition could not take place there on account of the velocity of the water." Wey then shows that a comparison of the gradients of the trench and of the Rhine channel just above its mouth indicates that the bottom current, while sufficient to prevent deposition, was too weak to produce the trench by erosion. The opinion of Wey and Forel was widely accepted, and effectively supplanted the earlier view, apparently originated by von Salis, that the trenches were erosional phenomena.

The conclusion that the sublacustrine trenches of the Rhine and Rhone are purely depositional phenomena becomes of vital importance when one comes to consider the problem of submarine canyons. In addition to the almost insuperable objections cited by Daly against considering submarine canyons as the product of differential deposition of the continental shelves, we may add three further objections: the enormous depth and breadth of the canyons under this hypothesis call for currents of titanic proportions, of which there is no evidence; the continental shelves do not exhibit the levee-like forms bordering the canyons which appear beside the Rhine and Rhone trenches, and which are expectable under the depositional hypothesis of canyon origin; submarine canyons have their greatest depths (below the adjacent shelf surface) far out from their heads, and not toward their inner ends as in sublacustrine trenches formed by differential deposition.

It should further be noted that the depositional hypothesis of Wey and Forel implies the existence of very weak bottom currents even when turbidity, salinity, and temperature combine to make the river water flow down the lake floor. If, as is generally admitted, the submarine canyons trenching the continental shelves can not be explained as phenomena of differential deposition, then under the turbidity-current hypothesis we must invoke turbidity currents powerful enough to plunge to enormous depths and to erode well-lithified deposits, even when salinity and temperature conditions are, as we shall see, opposed to such action. It is not surprising, therefore, that Forel was firmly of the opinion that there was nothing in common between the origin of the sublacustrine trenches he studied so ably and the origin of submarine canyons.

A most significant aspect of the problem presented by the Rhine and Rhone trenches, not covered by Daly's discussion, is the fact that these trenches are exceptional features. Forel recognized that other sediment-laden streams entering the Swiss lakes did not have trenches opposite their mouths, and that this presented a further problem requiring solution. As one element of the solution, he suggested[48] that trenches might be produced by differential deposition only where fine material carried in suspension greatly exceeded the quantity of coarse debris transported by the stream. "A heavy load of coarse gravel deposited by the inflowing stream immediately at its entrance into the lake would not permit establishment of the rather complicated mechanism of the sublacustrine trench." Delebecque[49] and Collet[50] later expressed similar views. Delebecque thought formation of the trenches was further dependent upon the river's having sufficient length for its coarse debris to be largely reduced by friction to fine material capable of being held in suspension and (following Schloesing[51]) upon the lake waters' containing a quantity of $CaO + MgO$ not less than 0.06 grams per liter, in order to insure rapid precipitation of suspended sediment. As noted on earlier pages, Heim and others recognized that turbid river waters sometimes spread over the surfaces of lakes; sometimes sink to moderate depths and then spread laterally between upper lighter and lower heavier masses of water; and sometimes sink to the bottoms of lakes, under which conditions alone are sublacustrine trenches formed. Even when sinking of heavier water below the lake surface is sufficiently sudden to produce a violent commotion readily visible to all, sublacustrine trenches are rarely formed. As Forel[52] pointed out, every Alpine river charged with glacial waters which enters a lake plunges violently under the lake waters; but sublacustrine trenches are few.

We are not here concerned with the question as to which investigators correctly diagnosed the reasons for the fact that turbid rivers do not always form trenches when they enter lakes. Until there is substantial agreement on this point we can not with any assurance apply their conclusions to conditions found in the ocean. But it is important for us to face the fact that whereas turbid rivers are numerous, subaqueous trenches off their mouths are comparatively rare; and to recognize that this implies balancing of diverse factors under such conditions that the presence of abundant sediment in water is not controlling as to the results produced. Indeed, as will be demonstrated below, muddy river water entering the ocean commonly floats on the heavier salt water, instead of sinking to the bottom as in lakes.

One further point deserves emphasis. Forel and other investigators have made clear the fact that where the sublacustrine trenches are found the sinking of the heavier river water is a spectacular phenomenon which can not escape popular attention. For nearly two thousand years, at least since the time of Pliny who commented on the frequent occurrence of the phenomenon, the plunging of heavy river water under lighter lake water has been a widely recognized curiosity of nature.[53] In Lake Constance where the heavy Rhine waters plunge downward the resulting commotion is popularly known as the *Brech*; in Lake Geneva the violent disturbance caused by the almost vertical descent of the heavy Rhone waters is called the *Bataillière*, "battle of waters." According to Forel the name is well deserved. "Small boats must employ caution in the violently agitated waves . . . . the whirlpool of waters at the surface [is] a gigantic gyration, a veritable Maëlstrom . . . . there is evident suction, which can only be explained by a vertical downward plunging of the water."[54]

If the supposedly weak density currents held responsible for the production, through differential deposition, of the insignificant trenches of the Rhine and Rhone give rise, at their inception, to such spectacular phenomena as the Brech and the Bataillière, what should we expect of analogous currents sufficiently vigorous and extensive to produce by mechanical erosion the gigantic gorges cut into the margins of the continental shelves? It seems reasonable to suppose that analogous currents would produce similar phenomena on a scale somewhat comparable to the greater results achieved. Nothing comparable is known where muddy rivers now enter the sea, nor where waves now beat upon muddy sea bottoms. Is it reasonable to suppose that when continental glaciers were more extensive than now, and sea level was a few hundred feet lower, conditions were so radically different as to produce on a grand scale phenomena now unknown along the continental borders? Or is it not more reasonable to believe with Forel that the explanation of submarine canyons must be found in some process wholly unlike that which produced the sublacustrine trenches of the Rhine and Rhone?

Daly closed his brief discussion of the work of Forel and his followers with the following statement: "The whole assemblage of facts won by the able, careful experts, Forel, Heim, Collet, and the hydrographers of the Swiss Government, seems strongly to favor the general conception of the genesis of the submarine trenches now being presented" (i.e., the conception of canyon cutting by turbidity currents). The writer would be inclined to close the present discussion with the statement that when the facts set forth by Forel, Heim, and their followers are critically examined they are found to offer no support for the turbidity-current hypothesis of canyon origin. Such support must be found, if at all, in evidence and arguments based on other foundation than analogy with trenches of the Rhine and Rhone rivers in Lakes Constance and Geneva. As Forel himself pointed out half a century ago, a satisfactory theory of origin for submarine canyons must be radically different from the theory advanced by him to explain sublacustrine trenches.

Currents in the Strait of Gibraltar. As a second analogy believed to support the turbidity-current hypothesis of submarine canyon origin, Daly cites the well-known double current at the Strait of Gibraltar. Here the upper waters to a depth of nearly 200 meters flow from the Atlantic into the Mediterranean basin at an average speed of four or five kilometers per hour, while the lower current flows outward to the Atlantic even more steadily with a velocity of the same order, which velocity rises to seven kilometers or more when wind and tide are favorable. "Yet the essential cause of each four-kilometer current is a difference of density between Mediterranean and Atlantic water that is only about 2/1000 of either. Here, then, we have a case of a bottom current powerful enough to move even gravel and yet caused by an

Submarine Currents

excess of density of the same order as that of sea water which is temporarily loaded with sediment."

The difference in density between the Atlantic and Mediterranean waters is primarily due to a difference in salinity, and the double current at the Strait of Gibraltar is commonly referred to as a salinity current. But one cannot safely attribute the observed normal velocities to the observed differences in density. The writer has elsewhere[55] described the Gibraltar currents at some length, and here quotes from that description, italicizing matter especially pertinent to the present discussion:

Evaporation is an effective agent in producing salinity currents, but in this case the surface current must of course flow inward toward the region of evaporation, where the water is increasing in density *and the surface is being lowered;* while the heavier salt water will flow outward at a lower level. A striking example of such circulation is found in the Strait of Gibraltar. The annual evaporation from the surface of the Mediterranean amounts to a layer of water at least 3 meters deep according to Fischer, and greatly exceeds the influx of fresh water, with the result that the waters in the sea become denser *and the surface lower* than is the case in the Atlantic Ocean. The *higher* and lighter waters of the Atlantic flow into the Mediterranean as a surface stream of marked strength, while deep-water observations prove that a strong current of more saline water moves outward on the bottom. The great velocity of these currents is a matter of considerable interest. Maury quotes the following from the abstract log of Lieutenant W. G. Temple for March 8, 1855, relating to the inflowing surface current: "Weather fine; made 1¼ pt. leeway. At noon, stood in to Almiria Bay, and anchored off the village of Roguetas. Found a great number of vessels waiting for a chance to get to the westward, and learned from them that at least a thousand sail are weatherbound between this and Gibraltar. Some of them have been so for six weeks, and have even got as far as Malaga, only to be swept back by the current. Indeed, no vessel had been able to get out into the Atlantic for three months past." It would seem from this that the surface salinity current, *reinforced no doubt by an hydraulic current due to heaping up of water in the Gulf of Cadiz under westerly winds,* and perhaps also to some extent by a direct wind current, had a velocity sufficiently great to prevent sailing vessels from passing westward to the Atlantic for months at a time. Helland-Hansen has shown that tidal currents also affect the movement of the waters in the strait, the direction of flow at a depth of 10 meters even being reversed from its usual inward course for a brief period on the day of his observations. . . . .

Investigations of the outlet of the Göta-Elf into the Kattegat showed that a reaction current flowed well into the bed of the river as a distinct bottom current of salt water. A sunken object was moved up the river channel by this current, in direct opposition to the surface flow. It was shown that this current could not be explained as a mere salinity current due to differences of specific gravity between the fresh and salt water. Ekman even goes further, and regards the bottom currents at the outlet of the Baltic Sea and in the Strait of Gibraltar *as in large part reaction currents*. Cronander, on the other hand, would seem to doubt the existence of true reaction currents, even at the outlet of the Göta-Elf where Ekman made his principal study. While *there are probably reaction currents developed* both at the mouth of the Baltic and *at the inlet to the Mediterranean*, Ekman seems to push his theory too far and to lose sight of the facts that salinity currents of large volume must exist under the conditions obtaining at such straits as those in question, and that any reaction currents found there are secondary phenomena of less importance than the currents which give rise to them.

It will be seen from the foregoing quotations that conditions at the Strait of Gibraltar are far from simple. The velocities of the inflowing and outflowing currents are not those due simply to differences in density, but result from a combination of favorable circumstances. Even were there no density differences between the Atlantic and Mediterranean waters, there would be a strong surface hydraulic current into the latter basin due to the lowering of the Mediterranean by excessive evaporation; a similar inward hydraulic current due to the action of prevailing westerly winds in piling up Atlantic waters in the Gulf of Cadiz; and presumably reaction currents outward along the bottom induced by the inflowing hydraulic currents. These currents would be independent of temporary wind and tidal currents which might accelerate them at some periods, retard or reverse them at other periods.

Even were the observed velocities of the Gibraltar currents due solely or primarily to salinity differences, they could not be considered as analogous to the turbidity currents invoked by Daly to carve canyons on an open continental shelf. The high velocities of currents in the Strait result from the funneling of water from vast areas through a comparatively narrow channel between two great land masses. There is no valid analogy between currents moving in the open sea and those flowing through a narrow opening in an extended land barrier. As is well known, tidal currents which are scarcely perceptible on the open continental shelf acquire very high velocities when forced through narrow inlets. The same must hold for currents of any origin.

Effect of Salinity on Deposition of Fine Sediment. One requirement of the turbidity-current hypothesis of submarine canyons is that sediment in large quantities must be held in suspension in sea water for long periods of time. This is necessary to enable the currents to flow outward over the shelf surface and far down its frontal slope without losing the load which is assumed to give the moving water its effective superior density. The well-known fact that fine sediment settles with comparative rapidity in salt water appears to create a difficulty for the hypothesis. Daly admits that the law holds for solid particles that are clayey and exceedingly small, but quotes Vernon-Harcourt's conclusion (in opposition to findings by Sidell) that river silts settle in sea water only about ten percent faster than in Thames River water, and cites Wheeler's experimental results to the same effect.

To evaluate the opinions of Vernon-Harcourt and Wheeler we must understand fully the particular problem they were studying and the nature of their investigations. In connection with studies of the Mississippi River, Sidell[56] endeavored to answer the question whether the river drops its load to form a bar or delta at its mouth solely because the river current is checked on meeting the ocean waters, or whether substances dissolved in sea water accelerate the precipitation. It thus appears that Sidell's attention was centered on the coarser debris deposited immediately at the river's mouth and not on the material carried in suspension far out to sea. His experiments, however, were such as to give a conclusion based on behavior of the finest material carried in suspension; for he found that Mississippi River water alone required from ten to fourteen days to settle (apparently meaning to become clear), whereas mixtures of such water with various salts and acids "became perfectly limpid in from fourteen to eighteen hours." Sidell's conclusion, that the coarser material forming the bar or delta was precipitated quickly because of substances dissolved in the sea water, did not properly follow from his experiments.

Sidell's demonstration that fine material in suspension is precipitated much more rapidly in salt water than in fresh was abundantly confirmed by later investigators; but his conclusion respecting the more rapid deposition of a bar or delta in salt water was frequently cited without realization of the fact that his experiments dealt with one type of sediment, his conclusion with another type. In 1868 Skey[57] showed that solutions of various neutral salts cause rapid clearing of water holding clay in suspension and suggested that "the transparency of the sea, into which is continually pouring such enormous quantities of turbid water, may be entirely due to the presence of so much saline matter." Schloesing[58] in 1870 secured similar results with clay, and also with the loam or silt of arable soils ("limons de terres arables"). He did not experiment directly with silts carried by rivers, but believed sea water should prove an active precipitant for such silts and hasten their deposition at river mouths. Three years later Waldie[59] found that during the rainy season the silt in the muddy Hugli settled very slowly. Basing his reasoning on the results secured by Skey and Schloesing, Waldie attributed this fact to excessive dilution, during heavy rains, of the saline matter normally found in the river water. By making good the deficiency of salines he brought about much more rapid precipitation of the mud. In 1883 Brewer[60] reported that clay from the Niobrara formation when added to pure water would not settle completely in thirty months; but that a like portion in salt water would settle completely in thirty minutes. Experiments with muddy waters of the Mississippi and Missouri Rivers gave similar results, and Brewer concluded that when muddy river waters enter the sea "practically it is the degree of saltness which controls the deposition." Further experiments[61] with fine soils and pulverized rock as well as with true clays confirmed Brewer's earlier conclusions. In 1886 Carl Barus[62] reported the results of experiments with fine solid particles of tripoli, red bolus, white bolus, talc, and bone ash. He found that all were precipitated more rapidly as the proportion of salt in the water was increased, and that the differences were readily apparent to the eye. Certain acids and alkalies are also effective precipitants. According to Barus and Schneider[63] "one molecule of HCl in ten thousand up to fifty thousand molecules of water is still clearly effective." J. Joly[64] in 1900 reported that the precipitating effects of salt in solution were marked in experiments with finely powdered carbon, kaolin, quartz, obsidian, and basalt.

It thus appears that by 1900 different investigators had firmly established the fact that very fine solid particles of many kinds of earthy materials (including clay, various fine river silts, silt of arable soils, tripoli, talc, quartz, obsidian, basalt, and other rocks) are precipitated much more quickly in salt water than in fresh water. They had shown, further, that the rapidity of precipitation increases with increase of salinity but not in the same proportion. Several of these investigators, and particularly many geologists not cited here, had without sufficient warrant invoked the results of these investigations to explain or help explain the deposition of bars and deltas at the mouths of rivers.

It was this last explanation against which Vernon-Harcourt in 1900 and Wheeler in 1901 directed their attacks. In so doing they, like those whose views they opposed, failed to make a sufficiently clear distinction between the various types of debris brought out by large and relatively sluggish muddy rivers to be deposited in the ocean: (a) Coarse sand and gravel moved along the stream bed largely by traction, and quickly dropped when the velocity of the current is checked, thus helping to form bars or deltas at the river's mouth. Samples of the river's water, even when taken from the lower part of the current, will contain little if any of this material. (b) Fine sand and relatively coarse silt, moved partly by traction and partly by saltation, but held in suspension only temporarily where eddying currents are active. While capable of being carried farther than coarse sand and gravel, this material is deposited with relative rapidity when the river's current is checked. Consequently it is the chief constituent of large deltas built by muddy rivers. Samples of the river's water taken near the surface of the stream will contain little if any of this material; but samples taken close to the stream bed will contain much of it. (c) Fine silt carried mostly in suspension. This material tends to remain in suspension so long as the river is in motion, and settles very slowly even when the river water becomes stagnant. As a result it is in large part carried past the delta, and scores or even hundreds of miles out to sea whenever the river water floats, as it usually does, on the heavier salt water of the ocean. Samples of the river water, even when taken at the surface of the stream, will contain considerable quantities of this material. (d) Extremely fine particles of clay and possibly other varieties of rock flour, carried in suspension, and capable of remaining in suspension in completely stagnant river water for days, months, or even years. Like the fine silt, this material is carried far out to sea, and samples taken from any part of the stream will contain appreciable quantities of it.

Of these four classes of material the first two settle so quickly in fresh water that addition of salts to the water cannot appreciably increase the rate of their precipitation. It is even conceivable that their precipitation might, under favorable circumstances, be slightly retarded, the cause* of the phenomenon described by Sidell, Skey, Schloesing, and their followers not operating effectively in the case of coarse debris, and the retarding effect of more dense saline water coming into play. It is materials of the two categories last mentioned above (c and d) that are supposed to remain long suspended in pure water, but to be quickly precipitated in salt water.

If with these fundamental facts clearly in mind we examine the work of Vernon-Harcourt and Wheeler,[65] we find that the results obtained by them, properly interpreted, do not invalidate the conclusion reached by earlier investigators to the effect that the presence of soluble salts in water greatly accelerates the precipitation of fine silt and other material carried in suspension. Vernon-Harcourt's experiments were directed to proving that bars and deltas at river mouths are due to checking of the river's current, and not to the action of sea water. In this he was doubtless substantially correct, for bars and deltas consist largely of materials too coarse to be affected by the presence of salts in water. It is vitally important to note that most of Vernon-Harcourt's samples, in part "deltaic silts," were taken from the bottoms of rivers where the coarser silt, sand, and other coarse debris are most abundant. Only one sample of truly suspended material was studied: mud from the Nile deposited in bottles filled near the surface of the river. And in this case the sediment was precipitated in thirteen minutes in sea water, but required three days to settle in distilled water, which even then showed a measure of turbidity. Vernon-Harcourt was not primarily interested in material carried in suspension, and remarked that the fine light matter carried in suspension in the upper layers of river water is not representative of the "silt" carried down by a river and eventually discharged into the sea. Elsewhere he states that with the exception of the Nile silt, there was no sign of that great difference in the rate of settlement of the samples of silt as a whole in sea water and in fresh water recorded by Sidell, "unless mere turbidity is taken into account." For Vernon-Harcourt, studying the origin of delta deposition, material in suspension and turbidity were properly negligible matters; and the study of deltaic deposits and bottom load of rivers was pertinent. But students of "suspension currents" or "turbidity currents" must not make

---

* The cause of the more rapid precipitation of fine sediment in saline waters is debated at length in some of the works cited and in other easily accessible literature, but is not directly concerned in the present problem.

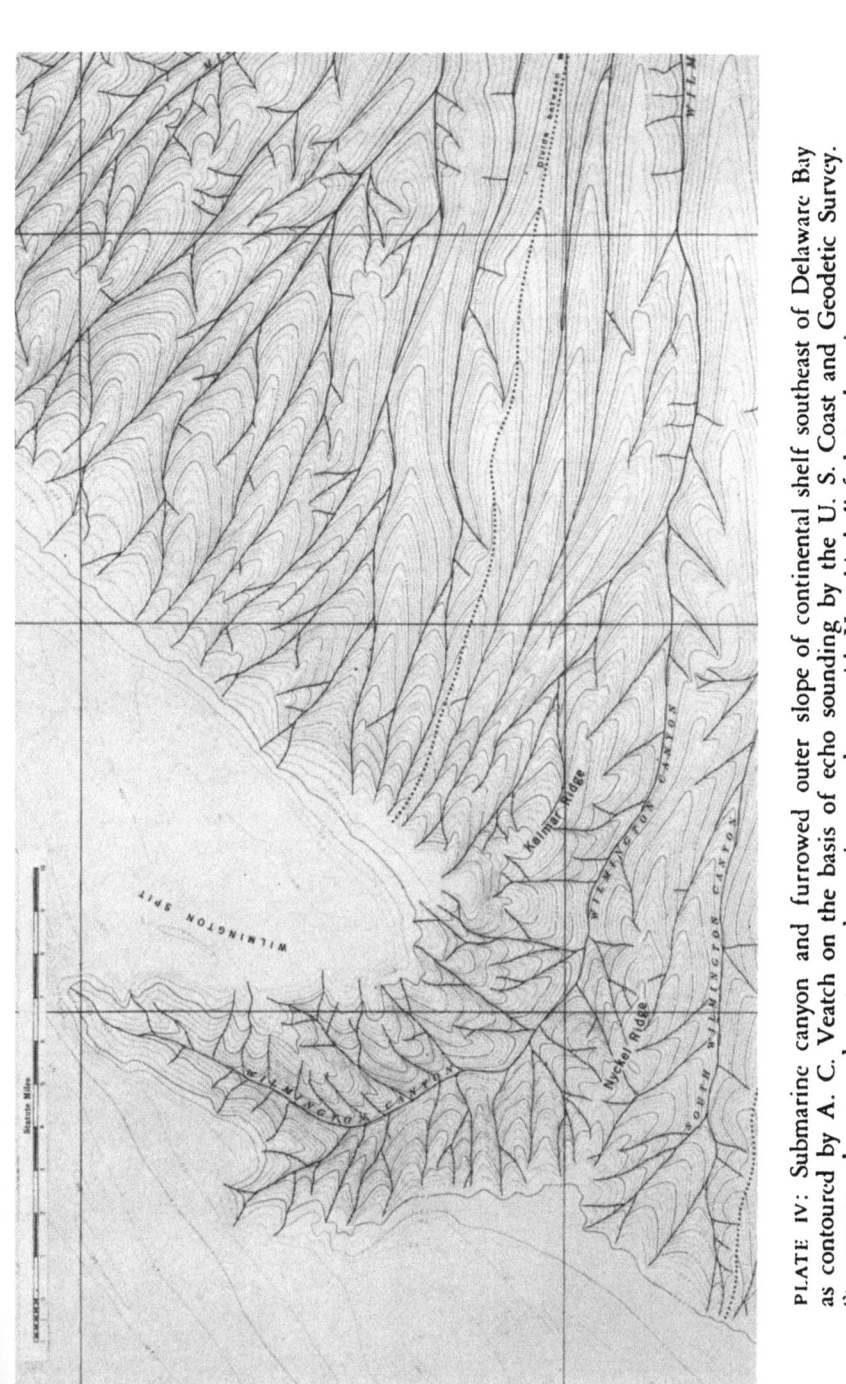

PLATE IV: Submarine canyon and furrowed outer slope of continental shelf southeast of Delaware Bay as contoured by A. C. Veatch on the basis of echo sounding by the U. S. Coast and Geodetic Survey. Streams are shown and contours drawn in accordance with Veatch's belief that submarine canyons represent submerged subaërial valleys. Contour interval: on shelf, 5 fathoms (30 feet) to depth of 75 fathoms; on slope, 25 fathoms (150 feet). Reproduced from Chart IIB, Special Paper No. 7, Geological Society of America.

the mistake of applying Vernon-Harcourt's conclusions to their problem.

In Vernon-Harcourt's experiments the river water employed was that of the Thames, but as this investigator points out, Thames water contains calcium bicarbonate and calcium sulphate in solution, both of which salts are active precipitants for fine sediment. As Vernon-Harcourt himself tells us: "These results furnish an explanation of the cause of silt settling in Thames water, with its small quantities of salts in solution, only a little more slowly than in sea water with its much larger proportion of salts in solution; for the chief salts in solution in Thames water are calcium bi-carbonate and calcium sulphate, very weak solutions of which have more influence in promoting the deposit of silt than the stronger solutions of the other salts experimented with, except potassium chloride."

Wheeler experimented with material deposited by rivers in flood plains and by brackish and salt-water currents in estuaries and salt marshes, rather than with material being carried in suspension by rivers. As Allen[66] acutely pointed out, Wheeler was dealing with material which had already undergone precipitation, and as he apparently took no precaution to remove metallic salts which might have caused precipitation the first time, these may have remained and prevented the particles from going into suspension when the experiments were made. For this reason Wheeler's results are not applicable to the problem of suspension or turbidity currents, although this investigator properly rejected Sidell's conclusion that precipitation by salt water was an important factor in bar and delta formation. That Wheeler seriously misunderstood some phases of the problem with which he was dealing is further indicated by his statement that "the very great distance over which solid matter brought down by rivers remains in suspension after reaching the sea, extending from six miles at the mouth of the Rhone to thirty-five from the outlet of the Nile, up to 300 miles over which the sea water is stated to be discoloured by the effluent of the Amazon, appears to indicate that salt water is capable of retaining solid matter in suspension for a longer time than fresh water." Apparently Wheeler was not familiar with the phenomenon of muddy river water floating for vast distances over the heavier salt water of the sea.

We conclude that salts in solution probably play a negligible rôle in causing deposition of the materials found in bars and deltas at the mouths of rivers entering the sea, partly because such salts can not greatly accelerate the deposition, rapid in any case, of relatively coarse silt and still coarser material transported by large rivers, and partly because the lighter fresh river water does not readily mix effectively with the heavier sea water; but that salts in solution are highly effective in precipitating fine silt and other fine material carried in suspension, when the salt solutions actually come in contact with such materials.

Since material of the first category [subdivided into (a) and (b) of the four types discussed on an earlier page] is quickly dropped close to shore when brought by rivers, and quickly redeposited when stirred up by wave agitation on the sea bottom, it is difficult to see how it can figure prominently as a factor in developing turbidity currents. And since material of the second category [subdivided into types (c) and (d) on an earlier page] is rapidly precipitated in the presence of soluble salts, and with difficulty taken into suspension again in the presence of such salts, it is hard to see how it can load sea water to a sufficient extent and for a sufficient time to produce turbidity currents of appreciable importance on the floor of the sea. The existence of fine sediment on outer portions of the continental shelf and down its frontal slope is eloquent testimony to the absence of effective currents of this type at the present time, and casts doubt on their existence in the Glacial epoch.

Degree of Lithification of Shelf Sediments. As demonstrated on earlier pages, under the turbidity-current hypothesis of submarine canyon origin currents of this type can not be assigned the lesser task of preventing deposition along their lines of flow while the continental shelves were built up on either side. They must be supposed to have eroded the canyons after the shelves of pre-Pleistocene deposits had been formed. It thus becomes important to enquire into the degree of lithification of sediments composing the shelves.

Daly[67] suggests that the degree of lithification of the shelf sediments may not be very great, and quotes Stetson to the effect that *"most of the material* of the [canyon] walls is not hardened into *rock* in the ordinary sense of that term," and that "the steepness of the walls at Georges Bank *need imply* rigidity of the material *in general* no greater than that illustrated in the visible clay of the Island of Martha's Vineyard, where the clay, like typical loess, is tenacious enough to stand in nearly vertical cliffs and yet offers little resistance to attack by running water."* Daly

---

* All italics are by the present writer.

also raises questions as to when the lithification was accomplished, and as to whether it may be concretionary and local.

As to the two questions last quoted: manifestly it is not easy to date with certainty the lithification of sediments in the submerged continental shelves. But no hypothesis of canyon origin can command confidence which depends upon the precarious assumption that Cretaceous and Tertiary sediments escaped lithification throughout the long period from their deposition to Glacial time, only to undergo lithification after the canyons were cut. Since all three canyons examined by Stetson on Georges Bank had lithified sediments in their walls, a canyon farther south revealed similar conditions, and "rock" has been reported from canyon walls in other regions, any assumption that lithification is concretionary in nature or local in extent must remain wholly gratuitous until proofs to this effect are forthcoming. The more reasonable assumption is that the materials repeatedly dredged from different canyon walls are fairly representative of shelf sediments in the areas examined.

This brings us to consideration of the degree of lithification indicated by materials thus far recovered from the canyon walls. Comparison with the clays of Martha's Vineyard seems of doubtful value. If the clays referred to by Daly are those exposed in the steep and picturesque cliffs of the classic Gay Head locality, it may be noted that they are intimately associated with layers of "puddingstone or conglomerate," "greensand marl much altered above by iron solutions," "coarse brown sand partly indurated into sandrock," "a heavy stratum of disintegrated rock in the condition of moderately loose sand," "black layers of lignite," "osseous conglomerate," "coarse sand . . . . sometimes indurated into a moderately coherent sandstone," and other beds which have helped to give a greater degree of rigidity to the section than is implied in the term "clays" commonly assigned to the mass as a whole. It will be noted, also, that there is evidence of the disintegration of beds formerly more resistant than now; and Uhler states that the beds "have been softened by atmospheric agencies."[68] Similar softening on exposure is reported by Shaler when describing the near-by Weyquosque clay beds. "Where the Weyquosque beds on Martha's Vineyard have been freshly bared by the frequent landslips which occur there the material is so firm that an ordinary pick can not well be driven into it at one stroke to the depth of more than one or two inches. After exposure to the action of the atmosphere the material becomes relatively very soft."[69]

The behavior of sediments on land which are close to the surface and free from a heavy overburden, which are subjected to mechanical disintegration by variations in temperature, by frost action, by alternate wetting and drying, and by other physical changes, and which are subjected to erosion by swiftly flowing concentrated streams of fresh water can afford no clue to the behavior of sediments encountered by submarine currents cutting deeply into a submerged continental shelf. A safer procedure is to ascertain the nature of the sediments as dredged from submarine canyon walls, and to judge as best we can whether materials of their character will yield readily to such currents as may reasonably be supposed to exist near the shelf margins.

Fortunately the investigations of Stetson and others are giving us a rapidly increasing amount of data on this point. In reporting the results of a dredging trip (1934) to the canyons cut in Georges Bank, Stetson[70] states that in places the slope of the canyon walls "is considerably above the angle of repose for unconsolidated material, *which indicates that these walls consist of rock.*"* Dredging brought up from one canyon wall "*a fairly well indurated sandstone*" of upper Cretaceous age, some fragments of which "showed weathered surfaces on all sides and were *obviously talus lying at the base of the cliff*; other pieces showed freshly broken faces." From another canyon wall "*a friable glauconitic sandstone*" of Navarro age was secured, and from a third "an impure glauconitic *sandstone,* late Tertiary in age." The walls of three canyons yielded a "*hard* green silt" or "*indurated* green silt" believed to be of late Tertiary age which "twice anchored the dredge firmly enough to check the ship's headway," and which "came up in large angular chunks, and was *sufficiently compact and firm* so that neither the washing it received on the trip to the surface, nor the strong compression in the dredge while towing, obliterated the freshly broken appearance of some of the faces." Stetson speaks of "the upper Cretaceous and the late Tertiary *rocks,*" and says "there can be no question but that the fossiliferous *rock* was found in place." Daly[71] quotes Stetson as stating verbally that "much less than half of the wall of any Georges Bank trench *is endowed with the strength of hard rock,* the lithified material occurring in individual layers separated by soft layers."

A later dredging expedition (1935) to the outer part of the Hudson

---

* All italics in these quotations are by the present writer.

submarine canyon, to two canyons cutting the shelf margin off the Delaware coast, and to one off Maryland showed Stetson[72] that "in many places the slope of the walls is apparently *above the angle of repose of unconsolidated material.*" Surface deposits were unconsolidated, but more compact clay and other material was dredged from the steep canyon walls. "No surface-deposits could ever attain the degree of compaction displayed here. In a tow from the southernmost valley the dredge came up full of fragments of a coarse-grained, *highly indurated sandstone* which unfortunately was unfossiliferous. *These were obviously talus blocks,* as they were weathered on all sides." In 1936 during further dredging in the submarine canyons of Georges Bank "two new *hard sandstone formations* were encountered in the canyon walls," one of these being identified as Peedee (upper Cretaceous) in age.[73] These last observations, and possibly those of the preceding summer, were made by Stetson subsequent to the verbal observations quoted by Daly and cited on an earlier page.

Taken altogether the degree of lithification described by Stetson appears highly unfavorable to the hypothesis of canyon cutting by turbidity currents operating at the margin of the continental shelf. Shepard's report[74] that twelve out of fifteen canyons examined by him off the coast of California "have definitely rocky material on the walls" is in accord with Stetson's findings. Thus the question is not whether turbidity currents during the Pleistocene were strong enough to erode "loose sand and mud." It is rather whether they were strong enough to erode sandstone and possibly other material "endowed with the strength of hard rock," even if such hard rock forms much less than half the canyon walls and occurs in individual beds separated by "indurated silt" and possibly by other material less resistant to erosion. To such a question one must hesitate to give an affirmative answer.

Having shown that evidence offered in favor of the turbidity-current hypothesis of canyon origin is less convincing than has been believed, we turn to other considerations which appear to militate against acceptance of this hypothesis.

Effects of Salinity and Temperature on Turbidity Currents in the Ocean.
Due to the admixture of fresh water coming from the land, sea water in the zone of wave agitation is less saline and therefore lighter than sea water deeper down. Thus salinity conditions in the ocean, unlike those

in the Swiss lakes previously discussed, are opposed to the operation of turbidity currents. It is true that the rivers of fresh water also bring vast quantities of sediment into the sea, and Daly appeals to them as one source of the material supposed to create turbidity currents on the continental shelves. But abundant observations indicate that where muddy river water enters the sea the material in suspension does not overcome the effects of salinity differences which tend to keep the fresh water on the surface, and that the muddy river water does not contribute effectively toward rendering the underlying sea water turbid. Instead of sinking, the highly turbid river waters float for vast distances over the more saline, denser sea water below, sometimes extending seaward beyond the margins of the continental shelf. Thus the muddy waters of the Nile, Ganges, Brahmaputra, and other rivers float seaward 40 to 100 miles from the coast.[75] The turbid waters of the Amazon spread eastward 200 miles, to and beyond the shelf margin.[76] The sediment-laden waters of the Mississippi, "floating over the heavier salt water, spread out into broad superficial sheets or layers, which the keels of vessels plough through, turning up a furrow of clear blue [salt] water."[77] For many miles off the mouth of the Congo "the water has a dark reddish-yellow colour, but this forms only a thin layer, as the ship's propeller turns up the colourless salt water beneath."[78] Sailing directions for this region issued by the U. S. Hydrographic Office,[79] point out that "a vessel is at times almost unmanageable" because a difference in current direction of the superficial muddy fresh water and the underlying clear salt water interferes with steering.

Whether the salt water could become charged with sediment settling from a muddy layer of overlying river water in sufficient quantity to initiate a descending current of turbid sea water seems highly doubtful. Very fine material settles far more rapidly in salt than in fresh water, while coarser silt will not remain long suspended in either when the waters move slowly and without turbulence. To what extent lighter land waters are mingled with heavier sea water by wave agitation, and the turbidity of the one is thus communicated to the other, is not known. In any case the mixed waters may, despite their turbidity, be lighter than the more saline sea water below. The fact that clear sea water is found immediately below turbid fresh water far off the mouths of large muddy rivers seems to indicate that a river's turbidity is not communicated to the sea water to any great extent. Near the mouth of a river where material of moderate coarseness is quickly precipitated and where wave turbulence in shallow water is at a maximum, there may be some mixing. But in these localities it is common to find reaction currents combined with salinity currents flowing *toward* the river's mouth along the bottom,[80] a fact seemingly inconsistent with the development of seaward-moving density currents at the place most favorable to them.

Unlike the Rhine and Rhone waters entering the Swiss lakes, sea water in the zone of wave action on those continental shelves cut by submarine canyons is always warmer than the deeper ocean waters. This gives the surface waters a further deficiency in density which must militate against their assumed descent into the depths. Even could turbidity overbalance at the surface the lightness due to lower salinity and higher temperature, and so start the turbid waters on their way down the submarine slope, it is difficult to see how such movement could fail to be quickly checked by the increasing salinity and lower temperatures encountered in depth, by the loss of turbidity due to deposition of load, and by frictional resistance against the shelf slope below and the wall of relatively stagnant water above.

Inadequacy of Turbidity Currents. If sediment-laden water flows down the upper surface of the continental shelf with sufficient velocity to arrive at the outer margin still so heavily charged as to be capable of descending the steep scarp to great depths with canyon-cutting power, one is led to ask whether there would be opportunity for fine sediment to accumulate on the shelf, and so be available for agitation by the waves of a glacially lowered sea. Would not a process supposedly so effective keep the shelf swept clean of fine material? If, on the other hand, we assume very slow movement down the surface of the shelf, with consequent opportunity for deposition of fine sediment en route, can we reasonably suppose that when the lightened waters reach the shelf margin they will still be so heavily charged with debris as to descend the steep scarp for thousands of feet into oceanic waters increasingly dense, and in so doing to develop velocities of canyon-cutting power? Such are the two horns of the dilemma presented by the basic assumptions of the turbidity-current hypothesis. Between those horns one is forced to seek a tenable position by assuming sufficient deposition to keep a supply of fine sediment on the outer borders of the shelf during pre-Glacial and inter-Glacial times; yet sufficient velocity for sediment-laden waters to travel the shorter distance (because of lowered sea level) to the shelf margin during Glacial time and arrive there still heavy enough to plunge effectively into the deeps. The dilemma must be faced, together with the fact that its existence, and the assumptions necessary to avoid its two horns, inevitably weaken confidence in the turbidity-current hypothesis of canyon origin.

Even if we assume that effective densities can be developed in the manner postulated, it seems more difficult to account for segregation of the resulting sheet of moving water into restricted threads of swifter flow than in the case of some other types of submarine currents. Wave attack on the surface of the nearly flat shelf would be distributed with exceptional uniformity along the coast. Appeal has been made to segregation by initial inequalities on the shelf surface due to inequalities of deposition when the shelf was formed. But such inequalities are numerous and of small magnitude, and give no such basis for assuming concentrated currents at widely spaced intervals as may be appealed to where reaction currents, salinity-density currents, temperature-density currents, tidal currents, and certain other types of currents are supposed to develop opposite widely separated river mouths, tidal inlets, or coastal embayments.

The magnitude of submarine canyons calls for a continuing process, enduring at least for long periods of time; whereas it is difficult to see how any hypothesis of turbidity currents can meet this requirement. Rivers constantly bring material in suspension to the sea, but we have seen that they are not effective in loading sea water with sediment. Testimony is abundant and consistent to the effect that where muddy river water floats over the heavier water of the ocean, the latter is found to be clear immediately below the turbid surface film. We can only conclude that sediment sinking from the muddy surface layer of fresh water is so quickly precipitated when it comes in contact with the salt water that the latter remains clear. Under these conditions it is doubtful whether even the bottom layer of sea water could become effectively loaded. In any case there is no satisfactory evidence that currents formed in this way are operating today, despite the continuing contributions of muddy water flowing into the oceans.

Recognizing that present hydrographic conditions cannot explain the canyons, Daly placed great weight on the effect of storm waves beating upon the continental shelf during the lowered sea level of Glacial time.

It is difficult, however, to see how the process here invoked can continue for any great length of time. Even if we assume that sedimentation during pre-Glacial and inter-Glacial times covered the outer part of the shelf with a greater proportion of fine silt than we find there today, and further assume the necessary competence of the currents, we face the difficulty of preserving the supply for the long period of time required to cut the deeper canyons. Waves are extremely effective in sorting sediment, and if the finer material is readily taken into suspension and swept over the shelf margin by turbidity currents, it would seem that the supply should quickly be exhausted. The rising and falling sea level would repeatedly shift the zone of maximum wave activity over a broad expanse of shelf; but only a part of this expanse, the outer and deeper parts, would have accumulated the finest material suitable for loading sea water effectively. It is, of course, difficult to evaluate the quantities of time and sediment involved in the equation, and conclusions must remain largely a matter of personal judgment. But the apparent difficulty of making the machinery of the turbidity current hypothesis operate continuously over significant periods of geologic time can not be ignored.

Flow of Turbid Water through Reservoirs. During the last few years an increasing number of observers have drawn attention to the fact that turbid water sometimes flows from a reservoir the upper waters of which remain clear, appearing as a muddy discharge through an outlet at or near the base of the dam.[81] Support has been sought in this phenomenon for the turbidity-current hypothesis of submarine canyon origin. The full significance of the fact is not yet apparent, and will not be known until highly important investigations now in progress are completed. Reference to this phase of the turbidity-current hypothesis with further discussion of possible velocity of such currents is presented by P. H. Kuenen.[82]

The hydraulic conditions existing in reservoirs, from which water is escaping through outlets at a significant distance below the surface, are not necessarily comparable with those existing in a lake with a surface outlet, or in the ocean with no outlet. The nature of the currents existing in such reservoirs and their effect upon the deposition of fine sediment (which may settle slowly in any case and never completely so long as there is even slight movement) remain to be determined. The possible effect of narrows, such as exist above Boulder Dam, in accelerating currents and causing turbidity may deserve study. One must further exclude the possibility that sediment slowly settling, as the water advances through the reservoir, leaves the surface portions clear but does not all reach the bottom before the outlet is attained. If this occurs, it proves the slowness with which sediment is deposited in slightly moving fresh water, rather than the vigorous flow of a restricted density current close to the bottom. These are but a few of many matters on which full information is prerequisite to any final opinion as to the real significance of the occasional issuing of turbid water from low reservoir outlets. Studies now in progress at Boulder Dam will doubtless throw valuable light on the whole question.

Grover and Howard, in discussing the outflow of turbid water from Lake Mead, state the problem in the following terms: "As the discharge of turbid water from a reservoir does not occur at all times when such water is flowing into it, there must be certain conditions of the inflowing water related to its specific gravity, viscosity, temperature, size, and degree of dispersion of the silt particles (or other physical qualities and combinations of them), that are prerequisite to such discharge. Although such discharges have been mentioned occasionally in engineering literature, the conditions under which they occur have not, as far as the writers are informed, been thoroughly described or studied."[83] O. A. Faris believes that the flows result from the disturbance of silt "which was virtually in place; that is, it was no longer in suspension, but had settled to the reservoir bottom soon after entering the slack-water and formed a 'flocculent' or 'honeycomb' structure of a consistency ranging from thick cream to heavy molasses."[84] The importance of flocculation in bringing about conditions essential to the flow of muddy water through reservoirs is stressed by a number of writers. In his report of the Committee on Density Currents for May 1, 1937, Eaton states that "the precise physical conditions that determine the occurrence and maintenance of such currents are not yet known. One of the important functions of this committee should be to decide what data are pertinent to the solution of the problem and to encourage the securing of these data at as many localities as possible. Density currents extending the entire length of a reservoir apparently occur at infrequent intervals; hence it may take years to obtain a satisfactory knowledge of them."

Pending the results of further study of the phenomenon of turbidity currents in reservoirs, judgment on the significance of the phenomenon as related to the problem of submarine canyons must be suspended.

Conclusion Respecting Turbidity Currents. We conclude our survey of the turbidity-current hypothesis with the realization that there are many obstacles, some of them of truly formidable proportions, in the way of its acceptance. As a working hypothesis it has repeatedly received attention during the last half-century. The theoretical objections urged against it, however formidable, do not justify us in removing it from the group of hypotheses deserving further study. It may in time find place, if not as the sole explanation of submarine canyons, perhaps as one factor in their creation, or at least in their preservation from obliteration by filling. But in the present limits of our knowledge the hypothesis appears of such doubtful validity that one is impelled to seek elsewhere a more satisfactory explanation of the great trenches found beneath the sea. Having failed to discover a wholly convincing explanation in subaërial and submarine processes, we turn to the remaining realm, the subterranean regions, to enquire whether within the solid earth, particularly within those parts projecting under the sea, there are processes in operation which offer a possible solution of the submarine canyon problem.

Notes and References

1. H. C. Stetson and J. Fred Smith, "Behavior of Suspension Currents and Mud Slides on the Continental Slope." *Amer. Jour. Sci.*, Vol. 35, pp. 1-13, 1938.
2. Reginald A. Daly, "Origin of Submarine 'Canyons.'" *Amer. Jour. Sci.*, Vol. 31, pp. 401-420, 1936.
3. W. E. Ritter, "A Summer's Dredging on the Coast of Southern California." *Science*, Vol. 15, pp. 55-65, 1902.
   W. S. T. Smith, "The Submarine Valleys of the California Coast." *Science*, Vol. 15, pp. 670-672, 1902.
   Wm. M. Davis, "Glacial Epochs of the Santa Monica Mountains, California." *Bull. Geol. Soc. Am.*, Vol. 44, pp. 1041-1133, 1933. See p. 1048.
4. J. Fergusson, "On Recent Changes in the Delta of the Ganges." *Jour. Geol. Soc. London*, Vol. 19, pp. 321-354, 1863. See p. 352.
   J. D. Dana, "Long Island Sound in the Quaternary Era, with Observations on the Submarine Hudson River Channel." *Amer. Jour. Sci.*, Vol. 40, pp. 425-437, 1890. See p. 432.
   Joseph LeConte, "Tertiary and Post-tertiary Changes of the Atlantic and Pacific Coasts; with a Note on the Mutual Relations of Land-Elevation and Ice-Accumulation during the Quaternary Period." *Bull. Geol. Soc. Am.*, Vol. 2, pp. 323-330, 1891. See p. 325.
   Bailey Willis, "A Submarine Trough off the Coast of Cyprus." *Geog. Jour.*, Vol. 79, pp. 349-351, 1932.

5. Wm. M. Davis, "Submarine Mock Valleys." *Geog. Rev.,* Vol. 24, pp. 297-308, 1934. See pp. 300-301.
6. Bailey Willis, *op. cit.*
M. J. de la Roche-Poncié, "Rapport sur la fosse et le havre de Cap-Breton (1860)." *Serv. Hydrog. de la Marine (France). Rech. Hydrog. sur le Régime des Côtes.* (Paris) Cahier 2 (1858-1863), pp. 62-74, 1877. See p. 69.
Wm. M. Davis, "Submarine Mock-Valleys." *Amer. Geophys. Union, Trans. 14th Ann. Meeting,* pp. 231-234, 1933. See p. 233.
7. J. Y. Buchanan, "On the Land Slopes Separating Continents and Ocean Basins, Especially Those on the West Coast of Africa." *Scot. Geog. Mag.,* Vol. 3, pp. 217-238, 1887.
8. F. L. Ekman, "On the General Causes of the Ocean-Currents." *Nova Acta Regiae Societatis Scientiarum Upsaliensis,* Series 3, Vol. 10, Art. 6, 52 pp., 1879. See p. 29.
9. For citations of the literature covering this phase of the subject see special discussion of turbidity currents on later pages.
10. H. C. Stetson, "Bed-Rock from the Continental Margin on Georges Bank." *Amer. Geophys. Union, Trans., 16th Ann. Meeting,* Part 1, pp. 226-228, 1935.
11. H. C. Stetson, "Current-Measurements in the Georges Bank Canyons." *Amer. Geophys. Union, Trans. 18th Ann. Meeting,* Part I, pp. 216-219, 1937.
12. Wm. M. Davis, "Submarine Mock Valleys." *Geog. Rev.,* Vol. 24, pp. 297-308, 1934. See p. 300.
13. Henry C. Stetson, "Geology and Paleontology of the Georges Bank Canyons," Part 1, *Geology. Bull. Geol. Soc. Am.,* Vol. 47, pp. 339-366, 1936.
14. Henry C. Stetson, *op. cit.*
15. Reginald A. Daly, "Origin of Submarine 'Canyons.'" *Amer. Jour. Sci.,* Vol. 31, pp. 401-420, 1936.
16. H. C. Stetson and J. Fred Smith, "Behavior of Suspension Currents and Mud Slides on the Continental Slope." *Amer. Jour. Sci.,* Vol. 35, pp. 1-13, 1938.
17. R. A. Daly, "Origin of Submarine 'Canyons.'" *Amer. Jour. Sci.,* Vol. 31, pp. 401-420, 1936.
18. Ad. von Salis, "Hydrotechnische Notizen: II. Die Tiefenmessungen im Bodensee." *Schweiz. Bauzeit.,* Vol. 3, No. 22, p. 127, 1884.
19. F. A. Forel, "Les ravins sous-lacustres des fleuves glaciaires." *Acad. Sci. Paris, Ct. Rend.,* Vol. 101, pp. 725-728, 1885.
20. F. A. Forel, "Le ravin sous-lacustre du Rhône dans le lac Léman." *Bull. Soc. Vaud. des Scis. Nat.,* Vol. 23, pp. 85-107, 1887.
21. F. A. Forel, *Le Léman.* Vol. 1. (Lausanne) 543 pp., 1892. See pp. 382-388.
22. *Ibid.,* p. 381.
23. A. Heilprin, "Geological Researches in Yucatan." *Acad. Nat. Sci. Philadelphia, Proc.,* Vol. 43, pp. 136-158, 1891.
24. Ernst Linhardt, "Ueber unterseeische Flussrinnen." *Jahresber. der Geog. Gesell. in München,* Vol. 14, pp. 21-52, 1892.
25. W. B. Scott, *An Introduction to Geology.* (New York) 2d ed., 816 pp., 1911. See pp. 140-141.
26. Wm. M. Davis, "Glacial Epochs of the Santa Monica Mountains, California." *Bull. Geol. Soc. Am.,* Vol. 44, pp. 1041-1133, 1933. See p. 1049.
27. Wm. M. Davis, "Submarine Mock-Valleys." *Amer. Geophys. Union, Trans. 14th Ann. Meeting,* pp. 231-234, 1933. See p. 233.
28. Wm. M. Davis, "Submarine Mock Valleys." *Geog. Rev.,* Vol. 24, pp. 297-308, 1934. See pp. 307-308.
29. Reginald A. Daly, "Origin of Submarine 'Canyons.'" *Amer. Jour. Sci.,* Vol. 31, pp. 401-420, 1936.
30. F. P. Shepard, "Daly's Submarine Canyon Hypothesis." *Amer. Jour. Sci.,* Vol. 33, pp. 369-379, 1937.

Notes and References

31. Paul A. Smith, Jr., "Submarine Valleys." *U. S. Coast and Geod. Surv., Field Eng.*, Bull. 10, pp. 150-155, 1936. See p. 154.
32. Paul A. Smith, Jr., "The Submarine Topography of Bogoslof." *Geog. Rev.*, Vol. 27, pp. 630-636, 1937. See p. 636.
33. Ph. H. Kuenen, "Experiments in Connection with Daly's Hypothesis on the Formation of Submarine Canyons." *Leidsche Geol. Mededeel.*, Vol. 8, pp. 327-351, 1937.
34. H. C. Stetson and J. Fred Smith, "Behavior of Suspension Currents and Mud Slides on the Continental Slope." *Amer. Jour. Sci.*, Vol. 35, pp. 1-13, 1938.
35. Reginald A. Daly, "Origin of Submarine 'Canyons.'" *Amer. Jour. Sci.*, Vol. 31, pp. 401-420, 1936.
36. F. A. Forel, "Le ravin sous-lacustre du Rhône dans le lac Léman." *Bull. Soc. Vaud. des Scis. Nat.*, Vol. 23, pp. 85-107, 1887.
37. Ad. von Salis, "Hydrotechnische Notizen: II. Die Tiefenmessungen im Bodensee." *Schweiz. Bauzeit.*, Vol. 3, No. 22, p. 127, 1884.
38. F. A. Forel, "Les ravins sous-lacustres des fleuves glaciaires." *Acad. Sci. Paris, Ct. Rend.*, Vol. 101, pp. 725-728, 1885.
39. F. A. Forel, "Sur l'inclinaison des couches isothermes dans les eaux profondes du lac Léman." *Acad. Sci. Paris, Ct. Rend.*, Vol. 102, pp. 712-714, 1886.

———"Le ravin sous-lacustre du Rhône dans le lac Léman." *Bull. Soc. Vaud. des Scis. Nat.*, Vol. 23, pp. 85-107, 1887.

———*Le Léman*. Vol. 1. (Lausanne) 543 pp., 1892. See pp. 373, 384, 386.

———*Le Léman*. Vol. 2. (Lausanne) 651 pp., 1895. See p. 275.

———*Handbuch der Seenkunde*. (Stuttgart) 249 pp., 1901. See p. 84.
40. F. A. Forel, "Le ravin sous-lacustre du Rhône dans le lac Léman." *Bull. Soc. Vaud. des Scis. Nat.*, Vol. 23, pp. 85-107, 1887. See pp. 95-99.
41. Eberhard, Graf von Zeppelin, "Der 'Bodensee-Forschungen' bezw. der Begleitworte dritter Abschnitt: Die hydrographischen Verhältnisse des Bodensees." *Ver. f. Gesch. des Bodensees u. sein. Umge.*, Sch., Vol. 22, pp. 59-103, 1893. See p. 81.
42. A. Delebecque, "Les ravins sous-lacustres des fleuves glaciaires." *Arch. des Scis. Phys. et Nat.*, 4e pér., Vol. 1, pp. 485-487, 1896.

———"Influence de la composition de l'eau des lacs sur la formation des ravins sous-lacustres." *Acad. Sci. Paris, Ct. Rend.*, Vol. 123, pp. 71-72, 1896.

——— *Les Lacs français*. (Paris) 436 pp., 1898.
43. Albert Heim, "Der Schlammabsatz am Grunde des Vierwaldstättersee." *Naturforsch. Gesell. Zürich, Vierteljahrssch.*, Vol. 45, pp. 164-182, 1900.

———*Geologie der Schweiz*. Vol. 1. (Leipzig) 704 pp., 1919. See pp. 430-431.
44. E. Kleinschmidt, "Beiträge zur Limnologie des Bodensees." *Ver. f. Gesch. des Bodensees u. sein. Umge.*, Sch., Vol. 49, pp. 34-69, 1921. See pp. 66-68.
45. W. Schmidle, "Die Geologie des Bodenseebeckens." *Ver. f. Gesch. des Bodensees u. sein. Umge.*, Sch., Vol. 50, pp. 38-55, 1922. See pp. 49, 50.
46. L. Du Parc, "Le ravin sous-lacustre du Rhône." *Arch. des Scis. Phys. et Nat.*, 3e pér., Vol. 27, pp. 350-353, 1892.
47. J. Wey, "Die Ungestaltung der Ausmündung des Rheins und der Bregenzer-Ach in den Bodensee während der letzten 20, bezw. 24 Jahre." *Schweiz. Bauzeit.*, Vol. 9, No. 6, pp. 36-37, 1887.
48. F. A. Forel, *Le Léman*. Vol. 1. (Lausanne) 543 pp., 1892. See pp. 383-384.
49. A. Delebecque, "Les ravins sous-lacustres des fleuves glaciaires." *Arch. des Scis. Phys. et Nat.*, 4e pér., Vol. 1, pp. 485-487, 1896.

———*Les Lacs français*. (Paris) 436 pp., 1898. See pp. 62-72.
50. Leon W. Collet, *Les Lacs*. (Paris) 320 pp., 1925.

———"Le charriage des alluvions dans certains cours d'eau de la Suisse." *Ann. der Schweiz. Landeshyd.*, Vol. 2, No. 1, pp. 1-192, 1916. See p. 151.
51. Ch. Schloesing, "Sur la précipitation des limons par des solutions salines très-étendues." *Acad. Sci. Paris, Ct. Rend.*, Vol. 70, pp. 1345-1348, 1870.

52. F. A. Forel, *Le Léman*. Vol. 1. (Lausanne) 543 pp., 1892. See p. 388.
53. *Ibid*.
54. *Ibid*., p. 386.
55. Douglas Johnson, *Shore Processes and Shoreline Development*. (New York) 584 pp., 1919. See pp. 134-135, 139.
56. A. A. Humphreys and H. L. Abbott, *Report on the Mississippi River*. (Bureau of Topographical Engineers, War Dept.), 1861. Appendix A (Survey by Capt. Talcott, 1838, Rept. of Assistant W. H. Sidell, pp. 5-14). See also: Reprint with additions, 1876, pp. 495-503.
57. Wm. Skey, "Coagulation and Precipitation of Clay by Neutral Salts Generally." *Chem. News*, Vol. 17, p. 160, 1868.
58. Ch. Schloesing, "Sur la précipitation des limons par des solutions salines très-étendues." *Acad. Sci. Paris, Ct. Rend.*, Vol. 70, pp. 1345-1348, 1870.
59. D. Waldie, "On the Muddy Water of the Húgli during the Rainy Season with Reference to Its Purification and to the Calcutta Water-Supply." *Proc. Asia. Soc. Beng.*, Vol. 42, pp. 175-178, 1873 (abstract). *Jour. Asia. Soc. Beng.*, Vol. 42, Part 2, pp. 210-226, 1873.
60. Wm. H. Brewer, "On the Subsidence of Particles in Liquids." *Nat. Acad. Sci., Mem.*, Vol. 2, pp. 165-175, 1883.
61. Wm. H. Brewer, "On the Suspension and Sedimentation of Clays." *Amer. Jour. Sci.*, Vol. 29, pp. 1-5, 1885.
62. Carl Barus, "Subsidence of Fine Solid Particles in Liquids." *U. S. Geol. Surv.*, Bull. 36, pp. 11-40, 1886.
63. C. Barus and E. A. Schneider, "Über die Natur der kolloidalen Lösungen." *Zeit. Phys. Chem.*, Vol. 8, pp. 278-298, 1891.
64. J. Joly, "On the Inner Mechanism of Sedimentation." *Royal Dublin Soc., Proc.*, Vol. 9, pp. 325-332, 1900.
65. L. F. Vernon-Harcourt, "Experimental Investigations on the Action of Sea Water in Accelerating the Deposit of River Silt and the Formation of Deltas." *Min. of Proc. Inst. Civ. Eng.*, Vol. 142, pp. 272-287, 1900.
W. H. Wheeler, "The Settlement of Solid Matter in Fresh and Salt Water." *Nature*, Vol. 64, pp. 181-182, 1901.
———*The Sea Coast*. (London) 361 pp., 1902. See pp. 62-65.
66. H. S. Allen, "The Settlement of Solid Matter in Fresh and Salt Water." *Nature*, Vol. 64, pp. 279-280, 1901.
67. R. A. Daly, "Origin of Submarine 'Canyons.'" *Amer. Jour. Sci.*, Vol. 31, pp. 401-420, 1936.
68. Charles Lyell, "On the Tertiary Strata of the Island of Martha's Vineyard in Massachusetts." *Geol. Soc. London, Proc.*, Vol. 4, pp. 31-33, 1843.
P. R. Uhler, "A Study of Gay Head, Martha's Vineyard." *Maryland Acad. Sci., Trans.*, Vol. 1, pp. 204-212, 1892.
———"Gay Head." *Science*, Vol. 20, pp. 176-177, 1892.
———"Observations on the Cretaceous at Gay Head." *Science*, Vol. 20, pp. 373-374, 1892.
69. N. S. Shaler, J. B. Woodworth, and C. F. Marbut, "Glacial Brick Clays of Rhode Island and Southeastern Massachusetts." *U.S. Geol. Surv.*, 17th Ann. Rept., Part 1, pp. 957-1004, 1896.
70. H. C. Stetson, "Bed-Rock from the Continental Margin on Georges Bank." *Amer. Geophys. Union, Trans. 16th Ann. Meeting*, Part 1, pp. 226-228, 1935.
———"Geology and Paleontology of the Georges Bank Canyons." Part 1. Geology. *Bull. Geol. Soc. Am.*, Vol. 47, pp. 339-366, 1936.
71. R. A. Daly, "Origin of Submarine 'Canyons.'" *Amer. Jour. Sci.*, Vol. 31, pp. 401-420, 1936. See p. 418.
72. H. C. Stetson, "Dredge-Samples from the Submarine Canyons between the Hud-

son Gorge and Chesapeake Bay." *Amer. Geophys. Union, Trans. 17th Ann. Meeting,* Part 1, pp. 223-225, 1936.
73. H. C. Stetson, "Further Investigations of the Submarine Valleys of Georges Bank." *Geol. Soc. Am., Proc. for 1936,* p. 105, 1937.
74. F. P. Shepard, "Daly's Submarine Canyon Hypothesis." *Amer. Jour. Sci.,* Vol. 33, pp. 369-379, 1937. See p. 371.
75. Charles Lyell, *Principles of Geology.* Vol. 1. 12th ed. (London) 655 pp., 1875. See pp. 425-427, 457, 472-474.
76. *Lippincott's Gazetter of the World.* Ed. by Angelo and Louis Heilprin. (Philadelphia) 2106 pp., 1931. See p. 57.
77. Charles Lyell, *A Second Visit to the United States of North America.* Vol. 2. (London) 385 pp., 1849. See p. 154.
———*Principles of Geology.* Vol. 1. 12th ed. (London) 655 pp., 1875. See pp. 425-427, 457, 472-474.
78. J. Y. Buchanan, "On the Land Slopes Separating Continents and Ocean Basins, Especially Those on the West Coast of Africa." *Scot. Geog. Mag.,* Vol. 3, pp. 217-238, 1887. See p. 223.
79. *Sailing Directions for the Southwest Coast of Africa from Cape Palmas to Cape of Good Hope.* 3d ed. U. S. Hydrographic Office, No. 105, 442 pp., 1932. See pp. 260-261.
80. J. Y. Buchanan, "On the Land Slopes Separating Continents and Ocean Basins, Especially Those on the West Coast of Africa." *Scot. Geog. Mag.,* Vol. 3, pp. 217-238, 1887. See p. 223.
Douglas Johnson, *Shore Processes and Shoreline Development.* (New York) 584 pp., 1919. See p. 138.
81. See, for example, L. R. Fiock, "Records of Silt Carried by the Rio Grande and Its Accumulation in Elephant Butte Reservoir." *Amer. Geophys. Union, Trans. 15th Ann. Meeting,* Part 2, pp. 468-473, 1934.
H. M. Eakin, "Silting of Reservoirs." *U.S. Dept. Agri.,* Tech. Bull. 524, pp. 1-127, 1936.
N. C. Grover and C. S. Howard, "The Passage of Turbid Water through Lake Mead." *Amer. Soc. Civ. Eng., Proc.,* Vol. 63, pp. 643-655, 1937.
O. A. Faris and others, "Discussion of 'The Passage of Turbid Water through Lake Mead.'" *Amer. Soc. Civ. Eng., Proc.,* Vol. 63, pp. 1208-1214, 1937.
W. P. Creager, the same, pp. 1405-1407.
D. M. Forester, the same, pp. 1602-1603.
J. C. Stevens, the same, pp. 1810-1812.
R. E. Redden, the same, Vol. 64, pp. 781-784, 1938.
H. N. Eaton, "Report of the Committee on Density Currents." *Nat. Res. Coun., Ann. Rept. Div. Geol. and Geog., 1936-1937,* Appendix M, 3 pp., 1937.
"Report of the Subcommittee on Application of Hydrodynamics to Problems of Geology." *Nat. Res. Coun., Rept. Interdiv. Comm. on Borderland Fields, 1937,* pp. 34-37, 1938.
H. N. Eaton, "Progress-Report of the National Research Council Interdivisional Committee on Density-Currents." *Amer. Geophys. Union, Trans. 19th Ann. Meeting.* Part 1, pp. 387-392, 1938.
82. P. H. Kuenen, "Density Currents in Connection with the Problem of Submarine Canyons." *Geol. Mag.,* Vol. 75, pp. 241-249, 1938. See also same author, "Onderzeesche Canyons." *Tijdschrift Kon. Ned. Aardrijkskundig Genootschap Amsterdam,* 2e Reeks, D1, 55, pp. 861-876, 1938.
83. *Loc. cit.,* p. 664.
84. *Loc. cit.,* p. 1208.

CHAPTER IV
Hypotheses of Subterranean Origin

1. SUBTERRANEAN RIVER OUTLETS

In early discussions of submarine canyons it was suggested that they might be produced where "subterranean rivers" had their outlets below sea level. This hypothesis was usually if not always advanced in a tentative manner, and without any attempt to explain just how such rivers could produce great canyons where they debouched into the ocean. Nearly eighty years ago De la Roche-Poncié[1] mentioned the existence of the subterranean river hypothesis (although employing the expression *submarine* river) in a form which attributed the "fosse de Cap-Breton" to lack of deposition in front of the outflowing stream while the alluvium, pushed to either side, built up the adjacent areas.

Milne[2] in 1897, discussing "submarine springs" emerging on the sea bottom, attributes these springs to the outflowing of underground streams. Referring to supposed examples of such rivers in Japan, New Zealand, and elsewhere, he writes: "The phenomenon . . . . seems best explained by the assumption that it [the water] comes from an ancient river-course roughly parallel with the adjacent river on the surface. . . . . Whatever may be the explanation of a stream practically beneath a stream, the fact remains that there are in the places cited, and possibly in very many others, very large bodies of subterranean water flowing seawards." The context makes plain that the existence of the supposed subterranean rivers was predicated on the known existence of artesian conditions in the areas cited, the artesian water being attributed to underground streams.

Two years later Benest[3] noted that rivers in certain regions terminate in lagoons (apparently separated from the sea by bars), and concluded that these rivers had gradually been diverted to subterranean courses with the result that their waters, issuing below sea level, discharged where the submarine gullies are now found. "Many subterranean rivers are supposed, with good reason, to exist. These have their outlets in some cases in the form of artesian wells." In 1902 W. S. T. Smith,[4] discussing submarine valleys off the California coast, wrote as follows:

> The emergence of subterranean streams might at least account, in some cases, for the absence of deposits in the valley heads and their nearness to the shore, if not for the formation of the valley as a whole. . . . . There are . . . . at intervals along the coast, deposits of loose materials extending to a considerable depth below sea level, and through these, underground waters, under sufficient head, might find a submarine outlet.

The foregoing citations are typical of others that might be recorded, but sufficiently illustrate the tentative and very generalized form in which the subterranean river hypothesis is advanced, as well as the frequent tendency to infer the existence of such rivers where artesian conditions were found. With increase in our knowledge of the magnitude of submarine canyons, their branching forms, and the character of their walls, and with wider understanding of what is implied in artesian flow, appeal to the subterranean river hypothesis of canyon origin has become rare.

That subterranean rivers do exist, especially in limestone regions, is well recognized; and that some of these do debouch below sea level is equally certain. But as a general explanation of submarine canyons the hypothesis has historic rather than immediate interest. It does not seem necessary to discuss in detail the weaknesses of a hypothesis generally recognized as inadequate.

2. FOUNDERING OF SUBTERRANEAN CAVERNS

Closely related to the hypothesis outlined in the preceding paragraphs is that which would explain some submarine canyons as produced by the caving in of the roof of a cavern or tunnel formed by a subterranean river. Benest in the paper cited above[5] expresses the opinion that "a vast number of caves and underground river outlets must still be totally unknown," and points out that broken cables may in part result from " 'caving' in of parts of the shell over cavities." Gorceix,[6] in discussing the origin of the famous "fosse de Cap-Breton," refers briefly to the hypothesis that a subterranean river of warm water, following a great fault or joint from the land out under the submerged margin of the continent, developed subterranean caverns in soluble beds. Foundering of the roofs of the caverns would then explain the local "deeps" or "sink-holes" existing along the axis of the particular submarine canyon in question. Gorceix spoke of this as a "wholly local theory," and in view of the great magnitude of the canyon and its "deeps" he considered detailed refutation of the hypothesis unnecessary.

The difficulties in the way of accepting the river-tunnel collapse hypothesis as a general explanation of submarine canyons are so obvious, and recourse to it has been so rare, that further discussion of its weaknesses does not seem to be required.

3. SOLUTION ALONG FAULTS BY UP-RISING SUBTERRANEAN WATERS

As noted in Section 2, appeal has been made to a subterranean river of warm water following a fault in the submerged border of the French coast and producing caverns in soluble beds which then collapsed to give the "fosse de Cap-Breton." Justifiable modifications of this hypothesis might conceivably make it more widely applicable. Thus it is not essential that the up-rising waters come from a "subterranean river"; or that the waters should be warm; or that caverns with roofs should be formed. Ordinary groundwater, percolating through permeable rocks and finding outlet to the ocean floor along faults or other fractures cutting through soluble beds, might dissolve such beds to give more or less linear depressions or submarine canyons. It does not seem possible, however, that the hypothesis, even in this broadened form, can satisfactorily account for most of the known submarine canyons. The branching and curving patterns of many canyons, their systematic relation to the shelf margin, the enormous depth of the major examples, and the absence in most cases of specific evidence of faulting seem to preclude acceptance of this hypothesis as an explanation widely applicable. So far as the writer is aware no one would now so employ it.

4. SOLUTION ALONG FAULTS BY DOWN-FILTERING MARINE WATERS

In a paper[7] earlier cited, and in another[8] published the same year, Gorceix sought to explain the "fosse de Cap-Breton" as the work of marine waters penetrating downward through protective beds along a series of fractures, to reach a band of gypsum, rock salt, and clays. This hypothesis is in part related to one discussed in Section 2. "These

Non-deposition along Faults

waters, by dissolving the salt and gypsum and disintegrating the clay, would give rise to a series of sinks or cave-ins arranged in a row and having markedly variable depth and extent, thus giving to all this submarine area the extraordinary aspect revealed by careful study of the soundings." "The aspect is *karstic*, absolutely."

Gorceix did not present his interpretation as a satisfactory explanation of submarine canyons in general. On the contrary, he offered it to account for specific facts observed in a single region under particular geological conditions known to favor the interpretation advanced, and "quite apart from any general theory." It seems highly improbable that the requisite conditions can exist wherever submarine canyons are found, and so far as the writer knows no one has adopted Gorceix's interpretation as a hypothesis of general application. For these reasons it seems unnecessary to discuss it further.

5. NON-DEPOSITION ALONG FAULTS DUE TO UP-RISING SUBTERRANEAN WATERS

In 1912 Dubalen[9] attempted to solve the mystery of the "fosse de Cap-Breton," and invented a hypothesis which in his opinion was "bien simple." The fosse being "directly in prolongation of the great [fault] line with thermal springs" known to exist on land, one had only to imagine the hot waters rising from the submerged seaward extension of the fault to account for the fact

that a trench so narrow (300 m.) and so deep at one point (377 m.) located off the mouth of a river (the former mouth of the Adour) carrying material to the sea, and in a shallow sea where violent currents transport sand from shoal areas, was not long ago filled with debris. . . . The presence of hot waters in the bottom of the trench, where the temperature of the sea water should be but 5° or 6°, would give rise to powerful density and hydrostatic currents; these currents are sufficient to disperse the sand which superficial marine currents constantly move into the trench.

In support of this interpretation Dubalen cited determinations of water temperatures in the trench which showed 12 degrees at a depth of 30 meters, and 29 degrees at a depth of 200 meters; and predicted that observations in the great deep of 377 meters would show still higher temperatures. Dubalen speaks of the hot waters as coming from a subterranean river following the fault, and apparently believed that the sea bottom was built up on either side of the zone along which currents generated by the rising hot waters prevented deposition.

Gorceix[10] expressed doubt as to the validity of this interpretation, and stressed the desirability of having more extensive and better controlled determinations of water temperatures within the trench. Charcot[11] quickly replied by pointing out that in June of 1913 and July 1914 the *Pourquoi Pas* had made precise temperature observations in the region in question, some of them at the points indicated by Dubalen, and had in every case found a regular and normal decrease in temperature as the bottom was approached. The abnormally high temperature of 29 degrees at a depth of 200 meters cited by Dubalen was to be ascribed, in Charcot's opinion, to use of a defective thermometer or to employment of a method insufficiently exact.

It should be pointed out that Charcot's observations do not of necessity prove the invalidity of those cited by Dubalen. Warm waters rising at separate points along a fissure on the ocean floor may reach the surface without spreading widely if the waters are relatively quiet; or, if the waters are sufficiently agitated, may be so widely dispersed as to mingle intimately with the colder sea water. In the first case one investigator might have the good fortune to secure an observation directly in one of the columns of ascending warm water, whereas other investigators might fail to find it, although it was very near. Observations made under the second conditions specified (when waters are disturbed) would everywhere show normal temperature gradients, except at the bottom and close to the points of issue.

One may more readily hesitate to accept Dubalen's interpretation because the magnitude of the result produced seems out of proportion to the agency invoked. It is difficult to imagine such an outpouring of hot waters along a line many miles in length as would produce uprising currents sufficiently strong to prevent effective deposition of sand in a zone relatively broad and deep. Even if one accepts such explanation for the "fosse de Cap-Breton," its general application to submarine canyons will doubtless appear to most students of these forms as without adequate justification. As no one has proposed such general application, further discussion seems unnecessary.

6. SUBMARINE SPRING SAPPING

So far as the writer has discovered no one has attempted to explain submarine canyons in shelf margins as the result of long-continued sapping by submarine springs fed not by subterranean rivers but by waters, chiefly artesian, migrating through the sediments of the continental shelf to appear on its steeper seaward face. While such an explanation may appear incredible on first statement, we are not justified in failing to explore any new line of enquiry which offers even faint hope of success. As Stetson and Smith[12] have truly said, "when a situation arises wherein all the theories advanced for a major problem such as this [the origin of submarine canyons] encounter serious difficulties at some stage, no possible contributing factor can be ignored."

It was with this same conviction that the writer two years ago suggested, without attempting to discuss it, the hypothesis of canyon origin now to be examined more fully. The occasion was a communication on submarine surveying and physiography presented before the Committee on Continental and Oceanic Structure at the Edinburgh meeting of the International Union of Geodesy and Geophysics, and the substance of the hypothesis was stated in the following terms:

It is known that the coastal plain deposits carry water under heavy pressure out under the sea, and that such water rises in artesian wells drilled on islands or sandbars several miles off the coast. I have wondered whether it could be possible that some pervious bed or beds of the coastal plain, at least in occasional places, carry water under pressure to the edge of the continental shelf. If deep submarine springs should develop there, would not such springs perforce migrate backward into the shelf deposits, leaving canyons the depth of which would depend upon the depth at which the upwelling waters escaped on the face of the continental scarp? It will be recalled that such impressive features as the deep and long "alcoves" eroded in the scarps of lava plateaus in the northwestern United States have been ascribed to just such headward migrating spring action. Submarine canyons cut in continental shelves at present submerged off rocky coasts may have been carved when parts of the shelf were above sea level and served to take in water which then migrated down the dip to the scarp face; or aquifers in the older rocks may connect under ground with overlying pervious formations of the blanketing shelf deposits.[13]

In examining the submarine spring-sapping hypothesis it will be profitable to enquire (A) as to the possibility of spring development on the relatively steep outer slope of the continental shelf; (B) as to whether such springs, if they exist, are competent to excavate canyons of the magnitude observed; and (C) as to whether the forms and distribution of observed canyons are compatible with the spring-sapping interpretation.

A. SPRING DEVELOPMENT ON THE CONTINENTAL SLOPE

Submarine Springs in Shallow and Deep Water. If submarine springs of fresh water break out on the ocean floor, it is obvious that only those in shallow water will readily be detected. The waters from deeply submerged springs, assuming that they rise, will be too widely dissipated before reaching the surface of the sea to create any disturbance there, or to give differences in temperature or salinity sufficiently great to attract immediate attention. Only by happy chance could one hit upon the orifice of a spring deeply submerged, and prove its existence by securing a sample of water fresh or nearly so; or secure a temperature reading at or near the submerged exit of such character as to leave little doubt of the spring's existence. Under normal conditions we must expect that if springs break out upon the ocean floor, only those relatively close to the surface will be found by the investigator. Springs on the outer face of the continental slope might be numerous and yet remain wholly unknown.

Submarine springs of fresh water debouching on the sea floor are frequently observed where their orifices are not deeply submerged. Over thirty years ago Hitchcock[14] described a number of examples, some of them previously noted by Le Conte, from the Hawaiian Islands, where the existence of artesian conditions in seaward-sloping beds suggested that submarine springs should exist off-shore. He also cites the now well-known springs, likewise associated with artesian conditions, found from one to several miles off the coast of Florida; and mentions other examples off the mouth of the Mississippi River, the island of Cuba, and elsewhere. Since the time when Hitchcock wrote "little is known about them," submarine springs have been so abundantly observed off so many coasts that specific citations are no longer required. The phenomenon of fresh-water springs rising from shallow sea bottoms within a few miles of shore belongs in the category of well-established fact.

Submarine springs at greater distances from land are, for reasons already cited, less well known. Copious springs of fresh water welling up on small islands or keys off mainland coasts have frequently been cited as evidence that fresh water from the mainland passes beneath the sea to emerge as springs farther out, usually attracting attention only when appearing on islands or close to their shores. The existence of such springs where readily detected suggests that many more occur on less accessible parts of the sea floor. In a paper on artesian waters in Florida, Stringfield[15] states that "a large spring is reported about 16 miles off the coast east of a point about midway between Coronado Beach, in Volusia County, and Canaveral, in Brevard County." As noted on a later page, submarine fresh-water springs are reported from 15 to 20 miles off the coast opposite Charleston, South Carolina. Élisée Reclus[16] reports that among the reefs and islets of the Jardines, 25 to 40 miles south of Cuba, "springs of fresh water bubble up from the deep, flowing probably in subterranean galleries from the mainland" (Cuba). Hitchcock[17] cites Rowan and Ramsey's popular volume on *The Island of Cuba* as authority for the statement "that the [fresh] water is often forced by hydrostatic pressure to the surface far out at sea."

The idea that springs of considerable magnitude may exist in great depths, and even on the steeper seaward slope of the continental shelf, has been entertained by a number of writers. By some the springs were visualized as the outlets of subterranean rivers, as in the papers of De la Roche-Poncié, Milne, and Dubalen cited above. Benest[18] after citing Professor (Sir Edgeworth) David on "powerful springs of fresh water at Port Macdonnell [close to the Queensland coast, Australia] rising up from the floor of the ocean, and discolouring the water for some distance around," continues (basing chiefly on an account by Reclus): "The aspect of the surface waters goes to show that from some cause, probably artesian, considerable disturbances take place on the bed of the sea along the Coromandel, Ceylon, and Malabar coasts. At several points stretches of muddy water, coloured yellow or red, have been seen, even in great depths." According to Mill,[19] in limestone regions land waters in part "ultimately well up through the salt water of the sea, sometimes from depths of 100 fathoms or more." Keilhack[20] states that there are numerous submarine springs with copious outflow off the coast of the Dalmatian karst, and cites one such spring off Cape St. Martin flowing at a depth of 700 meters. He then cites K. Hofeneder as authority for the statement that when a flood enters the Popovo polje, columns of turbid water under great hydrostatic pressure rise from the bottom of the Bay of Slano and so discolor the sea water that foreign ships take another course, assuming the muddy water to indicate a shallow bottom. Andrée[21] writes: "The fact that submarine springs flow out on the sea bottom, at depths of hundreds of meters, is discussed in detail on the earlier pages of Volume I." (Efforts to locate Volume I of this work have been unsuccessful. The libraries interrogated report that it has not yet been published.) Milne,[22] Benest,[23] and others assert that deep cable breaks not due to earthquakes occur in certain regions most frequently during or just after the season of heaviest rainfall on the adjacent coast, and suggest that the breaks result from landslides or cave-ins provoked by excessive outflows from deep submarine springs at those periods. According to Andrée: "The steep portions of the continental shelf may lose their support as a result of constant undermining by outflowing ground water, and be restored to a more stable slope by submarine landslides. That outflowing ground water does enter into the problem as a causal factor is shown by the periodic caving and cable breaks on the east African and South American coasts in the rainy season, which breaks ceased only when the cables were laid elsewhere."[24] Other students, without relating the phenomenon to seasonal precipitation, have attributed cable breaks in deep water to slumping and sliding provoked by submarine springs of fresh water.

Under the heading *Submarine Springs* Milne writes:

> Another cause tending to disturb, not simply the faces of delta formations, but the accumulations of loose materials covering the steeper slopes fringing the submarine plain which bounds most continents, is the not altogether hypothetical assumption of the existence of submarine springs, which in some instances at least are of marked magnitude. [And again:] Without attempting to multiply examples which show that fresh water from the land escapes beneath sea-level, from what we know about rainfall, its evaporation and absorption, and the geotectonic conditions governing the flow of underground waters, that much of this escapes beneath the sea on the fringes of plateaus surrounding continents and islands, is apparently a legitimate hypothesis.[25]

Artesian Conditions in the Continental Shelves. There can be little doubt that "geotectonic conditions" often favor the migration of fresh water far out under the submerged margins of the continents. Geological literature contains abundant references to the fact that on many coasts structural conditions favorable to seaward flow of artesian waters are observed right up to the shore, and must logically be assumed to persist farther out. Artesian wells located at the seaward margin of a coastal plain, and producing water from depths of one to several thousand feet below sea level, are presumptive evidence that the observed artesian conditions continue in that submerged portion of the coastal plain known as the continental shelf. Limited proof is available where artesian waters on off-shore bars and outlying islands tap aquifers at depths previously calculated. Unfortunately, such bars and islands are usually but a few miles at most off shore, and far from the shelf margin. Hence for the most part we must depend upon reasonable inferences. As Jack[26] wrote of the seaward-dipping major artesian horizon in Queensland, "I cannot deny that the Lower Cretaceous formation may crop out at the sea bottom still further out to sea, but the fact is not proven, and in the nature of things it is not likely to be proven."

From facts observed at or near the shore we may, however, draw inferences of much weight. Where artesian water from deep wells at the shore or on off-lying bars or islands is fresh, as is often the case, the fact is usually accepted as proof that the original salt-water content of the marine sediments has been expelled by the seaward flow of land waters. When such deep wells continue to flow indefinitely without becoming salt, there is indicated a continuous seaward movement of land water through aquifers leading to exits some distance farther out, although not necessarily at the edge of the continental shelf. In judging how far seaward land waters may push the process of expelling salt water originally included in the shelf sediments during deposition, we should remember that under the sea, just as on land, artesian waters normally move more readily in the direction of the bedding, following the pervious layers, than they do across the bedding where they are likely to be checked by impervious formations. The great distance to the edge of the shelf should not unduly concern us, since we have instances in which a single artesian horizon has been traced and found productive many hundreds of miles from its surface outcrop.

In a report on groundwater resources of Sarasota County, Florida, Stringfield,[27] while speaking guardedly of conditions concealed beneath the Gulf of Mexico, definitely recognizes the possibility, apparently even the probability, that artesian water-bearing formations (Tampa limestone and Ocala limestone) of the mainland outcrop on the steep seaward slope of the continental shelf 125 miles from shore:

In Sarasota County the Tampa formation lies about 500 feet below the surface, and the Ocala still deeper. To judge from the available information in regard to the structure of these formations and the configuration of the bottom of the Gulf of Mexico, the outcrop of the Tampa formation is fully 125 miles off shore. [And again:] The structure of the Tampa and Ocala limestones beneath the Gulf of Mexico is, of course, not known. If it is that inferred by Cooke and Mossom . . . the top of the Tampa limestone crops out about 125 miles off shore, at a depth of more than 800 feet, and the top of the Ocala limestone crops out a short distance farther out, at a depth of more than 1,000 feet. It must be recognized, however, that the structure of these formations beneath the Gulf may not be as assumed and that they may crop out at less depth and nearer the shore.

That the structure assumed as most probable by Stringfield, Cooke, and Mossom could produce artesian springs on the seaward face of the shelf is shown by Stringfield with the aid of a hypothetical diagram here reproduced as Figure 1. Although drawn primarily to illustrate

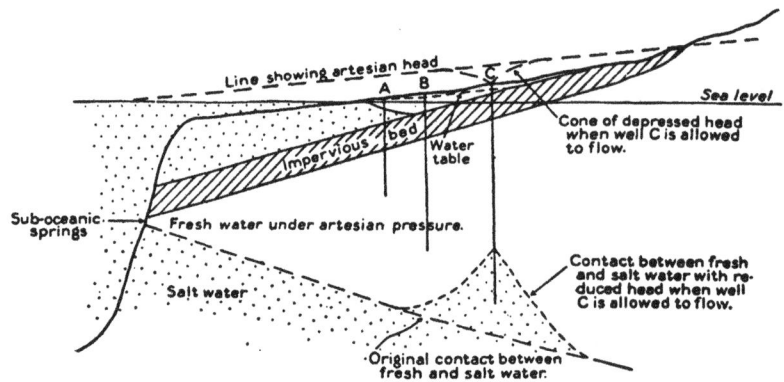

FIGURE 1: Diagram showing artesian conditions producing submarine springs on the seaward face of the continental shelf. Vertical scale greatly exaggerated. (After Stringfield, Florida Geological Survey.)

contamination of wells by salt water under varying conditions, it also shows that artesian water in a pervious formation beneath an impervious bed will appear as suboceanic springs if the artesian head is sufficient.

"If the head of water is sufficiently great a hydraulic gradient will be established in the water-bearing bed, the salt water will be pushed back to the submarine outcrop, and fresh water will escape into the sea."

Stringfield then shows that if the outlet spring were at the top of the submarine outcrop of the Tampa limestone at a depth of 800 feet, and if the specific gravity of the sea water were 1.025, the back pressure of the sea water would be sufficient to hold the fresh water in the limestone about 20 feet above sea level. The observed head at the shore is 24 feet. Since the excess of 4 feet is believed insufficient to cause flow of groundwater to an outcrop at a depth of 800 feet and 100 miles or more off shore, Stringfield concludes that the submarine outlet may be at some shallower depth, or nearer to the coast.

An early report[28] on artesian wells at Charleston, South Carolina, made by a selected scientific committee appointed to act in coöperation with the City Civil Engineer, is of interest in this connection. The report was based in part on stratigraphic and palaeontological determinations made by James Hall from samples submitted to him, and on other expert aid; and has been characterized by Darton "a model as a thorough and comprehensive document."[29] After describing artesian conditions of the region the report continues:

The Eocene and the Cretaceous formations encountered by our well, and others still deeper [1,945 to 2,000 feet], therefore probably continue their course [southeastward] under water for over seventy-five miles, when they are cut off by the deep submarine valley which forms the bed of the Gulf Stream. [As shown by context, the reference is to the steeper seaward face of the continental shelf.] There is evidence that the water contained in them finds a discharge into the sea. To this cause we must attribute the springs of fresh water that have been observed to rise, bubbling up at times in notable quantity, through the salt water at points along the coast, fifteen or twenty miles from the shore. In fact, we may well suppose that as these beds receive water at their upper outcrop against the granite ridge [Piedmont belt of crystallines], so they discharge it at their lower submarine outcrop along the side of the Gulf Stream valley [seaward face of continental shelf].

At the time of the report the artesian head at Charleston was comparatively high, 103 feet.

Dr. Arthur Howard[30] has called my attention to the fact that figures given for required heads, under different assumptions as to depth of fresh-water outflow, may need to be reduced where artesian waters increase their specific gravity by taking much mineral matter into solution. Conditions deep within the continental shelf, where both pressures and temperatures are presumably above normal, would seem to favor the solvent action of artesian waters. In his preliminary report on artesian wells in Georgia, McCallie, comparing the chemical composition of artesian waters and surface spring waters, writes:

The former, however, by reason of their high temperature and the great pressure under which they are confined, usually contain a much higher percentage of minerals, in solution, than the latter. . . . The most common solids found in artesian waters are the various carbonates, sulphates and chlorides, together with silica, alumina and iron. These several compounds, when present in unusual quantities, give rise to chalybeate, saline, magnesian and other mineral waters, many of which possess medicinal properties.[31]

In the records of deep artesian wells one frequently finds the water characterized by such expressions as "mineral water," "milky with lime," "chalybeate," "salty and alkaline," "very saline," "sulphur water," "white sulphur water," "contains iron," "magnesia water," "lithia water," "hot artesian salt water," "water very warm," etc. Where temperature records are given they frequently run from 70 degrees to 80 degrees or 90 degrees F., with occasional records up to 100 degrees and even higher. While dissolved minerals may appreciably increase the specific gravity of artesian waters, rise in temperature operates in the opposite direction.

It seems probable that the acquired content of dissolved minerals may under favorable conditions be sufficiently high to decrease noticeably the head required to cause suboceanic outflow of artesian waters. If such waters gain access to beds of rock salt or other sources of sodium chloride, they may acquire specific gravities far higher than that of sea water. While sodium carbonate does not usually occur in large amounts, it may occasionally be the chief mineral matter present, and may even be present in large amounts. In water from an artesian well 2,000 feet deep at Charleston, South Carolina, sodium carbonate is calculated as constituting more than two-thirds of the mineral content of the water, and as being more than four times as abundant as sodium chloride. Certain deep wells in Georgia, Missouri, and elsewhere have very much higher percentages of this readily soluble compound. "Magnesia" and "soda" in a "mineral" well 1,500 feet deep at Brunswick, Missouri, are present in amounts of 12,514 and 6,332 parts per million respectively, if the figures of the unknown analyst are correct. Many artesian waters in New South Wales are rich in sodium carbonate. Sodium sulphate is so abundant in artesian water from some parts of the Dakota sandstone as to be unfit for irrigation purposes.[32] Other mineral matter may be present in notable amounts.

It is not yet established, however, that the content of dissolved mineral matter in artesian waters plays a major rôle in the problem here considered. Figures on the solubility of most minerals found in such waters, and on the rise in specific gravity of the resulting solutions, indicate that while this factor can not be neglected, neither should its importance be exaggerated.

Where salt water originally included with marine sediments during deposition has not been completely expelled, the average specific gravity of waters in an artesian aquifer may be abnormally high, and the head required to produce flow correspondingly low. This fact may be of significance in the case of recent marine sediments through which artesian waters have moved for a comparatively short period. But it can hardly be a matter of major importance where long-continued flow of fresh water has more or less completely expelled connate waters from the sediments, or in the case of those parts of the shelf composed of continental deposits.

More significant is the possible effect of continuous introduction of sea water into a fresh-water artesian horizon. Dr. J. S. Brown[33] has pointed out that if the fresh-water head in an artesian horizon outcropping beneath sea level is depleted, sea water may gain access to the aquifer. This could readily occur by seepage downward through pervious beds or along fissures. There are, indeed, numerous references in the literature to leakage into and out of artesian water-bearing formations. As an example we may note the statement by Stringfield, in the report cited above,[34] to the effect that salt water from the Gulf of Mexico may have access to the Tampa and Ocala artesian aquifers, at less depth than their deep outcrops, "through permeable zones in the [overlying] Hawthorne formation or through openings in the confining beds of this formation."

Brown observes that if "salt water obtains ingress at an elevation higher than the more or less static fresh water the latter can then be expelled with a good deal of intermixture of salt water from strata at almost any depth." This is a mechanism which in Brown's opinion often operates "in complex ways in nature," and one which would seem to make possible extensive outflow of mixed fresh and salt water at great depths and far from land, even when the observed hydraulic head appears inadequate.

One may suppose, for example, that artesian water with a head H (Figure 2) escapes in part (dotted arrows) across adjacent pervious

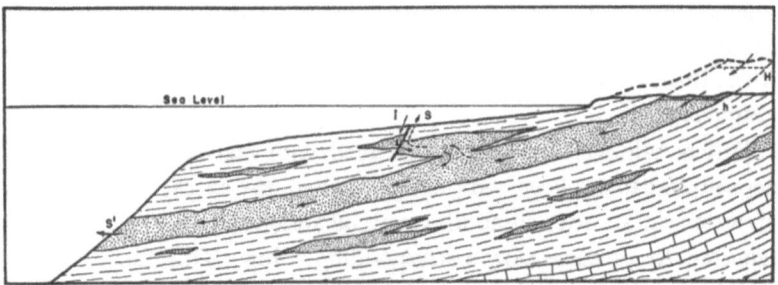

FIGURE 2: Artesian conditions permitting leakage out of (dotted arrows) or into an aquifer according as the head is high (H) or low (h). Vertical scale greatly exaggerated.

beds and through fissures or other openings to form a shallow submarine spring at S; and in part as deep submarine springs at S'. When erosion reduces the land surface, or subsidence lowers it, or rise of sea level occurs, the lesser head h may be insufficient to cause outflow. Under the new conditions, salt water flowing in through the fissure, now serving as an intake, I, will mix with fresh water in the aquifer, and the mingled waters will find exit at S'. It should be borne in mind that while fresh water descending along a fissure or through pervious beds will not intermingle freely with heavier salt water below, but rather will "float" upon it, salt water penetrating downward in the same manner tends, because of its higher specific gravity, to enter fresh water below and spread through it, increasing the weight of the mixture.

In discussing the observed low head (20 to 25 feet) originally found at Atlantic City in artesian wells approximately 800 feet deep Thompson[35] offers as one of three possible explanations the statement: "The salt water creating the head on the [water-bearing] sand may gain access to it much nearer shore than the edge of the continental shelf through

Submarine Spring Sapping

relatively permeable overlying materials, as in Figure 21 C." The figure referred to represents conditions similar to those in Figure 2 of the present volume, and bears the legend "Saltwater reaching [water-bearing] formation through break in overlying impervious clay." It is evident Thompson recognizes the fact that low artesian head may result when salt water gains access to the aquifer either through relatively permeable overlying beds, or through fissures or other breaks in overlying impervious deposits. He does not, however, discuss the effect of such conditions in permitting deep outflow of mixed fresh and salt water on the seaward face of the continental shelf even when the observed head is low.

Enough has been written to show that the escape of artesian waters far down the face of the continental shelf under existing geological and geographical conditions is by no means an impossibility. So far from such escape being incredible, it has repeatedly been considered as possible, or even highly probable, by competent students of the problem. In order to secure an expert opinion based on full acquaintance with modern developments in the field of groundwater behavior, the writer addressed to Dr. O. E. Meinzer, Geologist in Charge of the Division of Ground Water of the United States Geological Survey, two questions to which he kindly furnished the following answers:

> With reference to your question relative to artesian conditions, it appears reasonable that under favorable conditions water under artesian pressure may extend to the edge of the Continental Shelf in parts of the Atlantic Coastal Plain. Under such conditions the water will pass out through the submarine outcrop of the artesian formation, if the artesian pressure at the outcrop exceeds the back pressure caused by the ocean water.

After pointing out that the head must be sufficient to overcome pressure of the heavier sea water and also to produce flow through the aquifer for the required distance, Dr. Meinzer continues:

> The increase of the sodium chloride content of artesian water from wells nearer the coast does not necessarily indicate that there is no movement of water from the land to the edge of the Continental Shelf. Although in several parts of the coastal areas in Florida the artesian water is more highly mineralized near the coast than at some distance inland, the hydraulic gradient of the artesian water indicates movement of the water oceanward. This is true of most of the Atlantic Coast of Florida south of Saint Augustine.

Artesian Conditions in the Geologic Past. Thus far we have restricted our enquiry to artesian conditions existing at the present time and in the immediate past. But the hypothesis here under examination does not at all depend upon present or recent geological and geographical conditions. On the contrary, it recognizes that conditions during the Pleistocene and Tertiary, perhaps during part of the Cretaceous, and possibly even farther back in geologic time, may have been much more favorable to artesian outflow on the face of the submerged continental slope than is now the case.

Repeated lowering of sea level during successive stages of the Glacial period reduced the column of salt water which artesian outflow would have to displace. The proportionate reduction was not great if prevailing estimates of the amount of lowering (a few hundred feet at most) are correct. But it was of capital importance if some of the more radical estimates are nearer the truth. If, as some geologists believe, the land along the coast of eastern North America and certain other localities stood considerably higher in parts of the Pleistocene than now, as a result of upwarping, artesian waters may have circulated then where they fail to do so today. Dr. Meinzer has pointed out that "the large and sustained yield of wells in some places, as for example in the vicinity of Houston, shows that the formations have sufficient transmissibility to discharge large quantities of water if an adequate hydraulic gradient can be provided, as might have been the case in the Glacial stages."

Of even greater significance may have been the geological conditions of still earlier times. Artesian conditions off the Atlantic Coast are relatively unfavorable today. This is because erosion cycles preceding the present reduced non-resistant Coastal Plain formations to a coastal lowland but little above present sea level. Where the intake area of an artesian system is low the effective head cannot be high. Prior to the Somerville and Harrisburg erosion cycles the surface of the land at the eastern border of the Appalachians was hundreds of feet higher than now. In the Coastal Plain area the surface appears to have been lower than in the Appalachian belt, but higher than at present.

Prior to the Schooley cycle the Cretaceous beds presumably rose very much higher. The relatively steep angle at which the Fall Zone peneplane rises from beneath the Coastal Plain sediments, to be intersected by the later Schooley peneplane, implies a former higher elevation of the Coastal Plain formations west of their present westernmost outcrops. As has elsewhere been shown[36] there is very strong evidence to support the view that Coastal Plain formations extended westward upon the up-arched Fall Zone peneplane for 150 or 200 miles, passing above the present Appalachian summits (Figure 3). It should be noted that this view was developed before the present hypothesis of submarine canyon formation had been conceived, and hence before it was realized what strong support it would bring to this hypothesis.

The alternative view (rarely supported by geomorphologists) that down-wearing so nearly kept pace with uplift that outcrops of artesian horizons never attained a high elevation, while theoretically possible, is not a reasonable probability. In any event we must recognize the likelihood that outcrops of the most important artesian horizons may have been some thousands of feet above sea level throughout much of the long Schooley cycle. In such case the situation would have been comparable in essential respects to the artesian system of the Great Plains, where water enters the most important aquifers at elevations thousands of feet above sea level, flows under artesian conditions for "many hundreds of miles," and where tapped shows heads measured, not in a few tens of feet as along our present Atlantic coast, but in hundreds and even in thousands of feet.[37]

Not only were artesian conditions presumably far more favorable in the past than those observed today; they were doubtless also favorable in places where today they appear to be non-existent. When the land stood higher above sea level than at present, the weak Coastal Plain formations were in places entirely stripped from the underlying crystallines. Later submergence brought the sea against the crystalline basement. The crystallines of Connecticut descend beneath Long Island Sound, and it is only the presence of Cretaceous and Tertiary beds on Long Island which gives visible testimony to former geological conditions more favorable to artesian flow than those of the present. Off the coast of Maine the submerged depression of Long Island Sound is replaced by the broader and deeper Gulf of Maine, while Long Island finds its counterpart in the wholly submerged cuesta-shaped "banks." It surely is no unwarranted flight of the imagination to read in this submarine topography a close correlation with the cuesta and lowland topography characteristic of undrowned portions of the Coastal Plain farther southwest;[38] and to recognize that in the Maine region conditions must once have been far more favorable to artesian flow than the

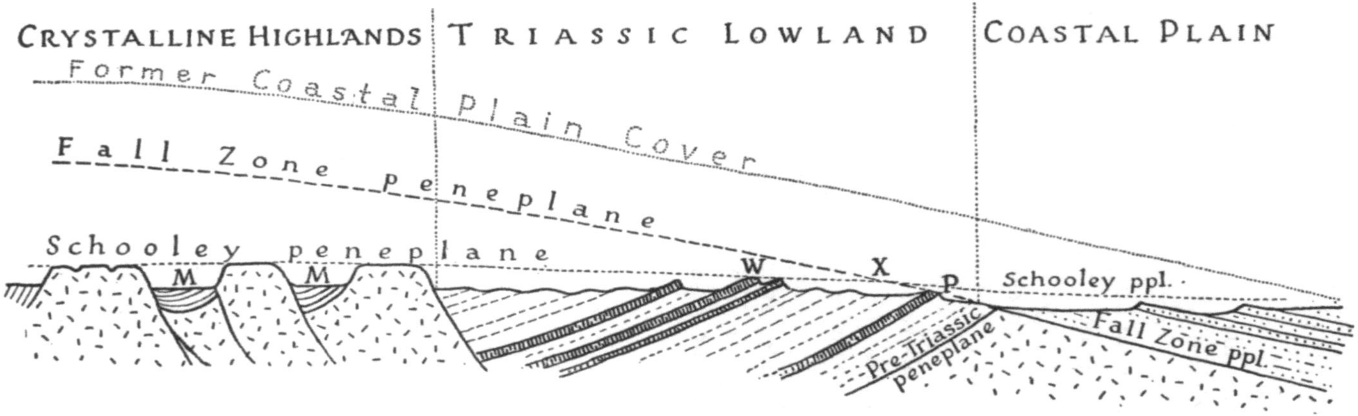

FIGURE 3: Portion of the Older Appalachians of northern New Jersey (at left), Coastal Plain remnant (at extreme right), and probable former continuation of the Coastal Plain beds high above the present Appalachian summits.

present geographical and geological features of the region would suggest. Geophysical studies reported by Bullard and Gaskell in the November 1938 issue of *Nature* indicate the presence of a series of sedimentary beds, from 2,000 to 4,000 or more feet in thickness, forming part of the continental shelf 115 to 170 miles off the coast of southwest England. Here, and off many another coast, the sea may hide structures which at some former time transmitted artesian waters to the shelf margin.

From the premises discussed above several important deductions may reasonably be made. The favorable geological conditions supposed to have existed in the past would have made possible artesian outflows at great depths, far down the seaward slopes of continental shelves, as well as at higher horizons. The lower formations of that part of the Cretaceous in the Atlantic Coastal Plain, for example, are among its most important artesian horizons. At what depths these outcrop on the shelf border, assuming they do so outcrop, is unknown. But outcrops 5,000 to 10,000 feet below sea level are possible. To produce an outflow at a depth of 10,000 feet there must be a fresh-water head of 250 feet to offset the back pressure of a 10,000-foot column of ocean water of average density 1.025, plus such additional head (say 1 foot per mile) as is required to produce flow of the artesian water from the outcrop on land to the submarine outlet. In the case of the Atlantic Coastal Plain of Schooley time this latter distance may have varied from 200 to 400 miles, making the total head required 450 feet for the smaller figure, 650 for the larger. As observed artesian heads sometimes amount to several thousand feet, and may well have been of this order of magnitude under the conditions believed to have existed in eastern North America during the Schooley cycle and possibly before, artesian waters may well have gushed from submarine outlets under very high pressure even at depths of 10,000 feet and more.

Such copious and vigorous outflow may have endured over a very great period of time: for parts of the Atlantic Coastal Plain, possibly from early or middle Cretaceous time onward, depending on a history not fully known; for other parts apparently from the close of the Cretaceous to far into the Tertiary. Even under the most moderate estimate the probable period of time during which vigorous artesian outflow was possible enormously exceeds the probable period of the lowered glacial sea level during which some investigators believe the submarine canyons were carved.

Conditions Affecting the Outcrop of Artesian Horizons. Whether artesian formations outcrop on seaward faces of continental shelves is, in the nature of the case, a question which can not be answered on the basis of positive direct evidence. The regions of the assumed outcrops are effectually concealed from observation. One might reason that if the outer parts of the shelves are deltaic in structure, with steeply plunging foresets inclined seaward, conditions would not favor submarine outcropping of aquifers. Or it might be argued that deposition of fine silt and clay on the face of a shelf would seal up such outcrops if they did exist; or that the water-bearing pervious formations may thin out and disappear before the seaward face of the shelf is reached.

To the first suggestion it must be answered that there is no satisfactory evidence to support the idea that the outer borders of existing continental shelves contain deltaic structure. If they do, such structure should not be visualized in terms of textbook diagrams showing steeply dipping foresets, or in terms of small deltas exposed on land and revealing foresets dipping at angles of 10 to 30 degrees or more. For it must be doubted, in the first place, whether typical delta structure (in so far as this term implies steeply dipping foresets making a large angle with nearly horizontal topset and bottomset beds) can be developed in the open sea, where wave and current action tend constantly to re-work coarse deposits in shallow and moderate depths, and to spread fine deposits widely and at low angles in deeper water. In the second place, observations on large deltas in the sea show low-angle frontal slopes, that of the Nile being 1.5 degrees according to Barrell.[39] Lawson[40] found the average frontal slopes of eight major deltas to be about one-half of one degree. Grabau[41] states that in large deltas of fine material "the topset beds may be more or less continuous with the fore-set, and, indeed, the two may imperceptibly grade into each other, without even a change in angle." According to Russell[42] the Mississippi delta lacks true foreset beds in the generally accepted sense of that term.

The so-called "steep" seaward slopes of continental shelves average less than one degree. As Salisbury says: "Even up most of these 'steep' slopes, therefore, railway trains could be run without change of grade."[43] Beds parallel to the slope would appear horizontal to the observer.

Submarine Spring Sapping

Nothing remotely resembling "deltaic structure" as ordinarily conceived is to be expected at the shelf borders.

Such evidence as we have points strongly to the conclusion that marine formations are normally non-deltaic in character. Much of the Atlantic Coastal Plain, and hence presumably much of the shelf which represents its submerged border, consists of continental deposits. Those portions of continental shelves now exposed as coastal plains, whether marine or continental in origin, show prevailing slight seaward dips. The uniformity of normal coastal plain structure from the Fall Line to the coast, a distance of 150 miles or more in places, with artesian formations operating effectively at depths of 2,000 feet or more, is presumptive evidence that such conditions persist indefinitely seaward. The fact that fresh water continues to flow from deep wells at the coast or on off-lying bars or islands, indicates that artesian waters at the depths tapped are flowing ever seaward. These waters must find their escape either (a) where pervious formations outcrop on the seaward-facing frontal slope of the continental shelf; or (b) across the bedding for thousands of feet to the upper surface of the shelf or across the bedding (possibly for a much shorter distance) to the frontal slope. As between the two latter routes, that to the frontal slope at great depths may be fully as favorable as that to the upper surface at lesser depths.

In any case, there is available at present no evidence which would justify us in excluding the possibility that artesian waters find exit at some places far down the seaward face of continental shelves; and hence no reason to deny consideration to a hypothesis of submarine canyon origin which invokes the action of outflowing artesian waters at great depth.

To the suggestion that deposition of fine silt and clay on the face of a shelf would seal outcrops of artesian horizons, it must be answered that unless the silts and clays are well consolidated, artesian waters under enormous pressure would saturate and eventually displace them; that once displaced they could never close the outlets so long as the artesian waters continued to flow; that silts and clays on the seaward face of a shelf are especially liable to slumping and sliding, even when fairly well consolidated; that such slumping and sliding has repeatedly been inferred as a potent cause of cable breaks, and might well expose the seaward ends of artesian aquifers; that continued deposition of silt and clay on the seaward face of a shelf is more likely to produce mudflowing (see later paragraphs) than to build up a thick deposit competent to seal the seaward sides of water-bearing formations. In any case we have no evidence to indicate that artesian horizons are in fact effectively sealed at all parts of their seaward borders.

That some water-bearing horizons thin out and disappear before reaching the seaward face of continental shelves seems highly probable. That other water-bearing horizons continue to that face seems equally probable. The persistence of artesian aquifers for hundreds of miles from their intake areas is a well-known phenomenon. Upon it depends the development of artesian wells over vast areas of the earth's surface. Just what is the origin of the relatively steep but in most cases absolutely gentle seaward-facing slopes of continental shelves, is unknown. If the nearly horizontal beds apparently forming much of the shelves bend gently downward at their seaward borders, they could nevertheless outcrop at successive levels on the slope. If the edges of the shelves represent fault scarps softened by slumping, the faults could easily cut any number of artesian horizons. If a former steeper depositional slope has been reduced to a low angle by slumping, such slumping may have exposed artesian aquifers at certain points. Under a variety of hypotheses outflow of artesian waters is conceivable. On the other hand, there are no known facts which exclude the possibility of artesian outflows on the seaward faces of continental shelves.

It should be borne in mind that the hypothesis of submarine canyon origin here under scrutiny does not demand continuous outcropping of artesian horizons all along the shelf margin. On the contrary, it specifically appeals to occasional outlets of major importance, fully recognizing that elsewhere outlets may have been minor or temporary in character, or wholly lacking. Thus it is quite permissible, under this hypothesis, to assume the closing of aquifers along most of the shelf border by any or all of the processes above discussed, *provided* that the competence of artesian waters to escape at some points far down the seaward slope of the shelf be granted.

Summary Respecting Artesian Conditions. If the reasoning set forth on preceding pages is valid, we are justified in closing this first part of our discussion of the hypothesis of shelf sapping by submerged artesian springs with certain conclusions not commonly emphasized: (a) Submarine fresh-water springs, abundantly developed in shallow waters bordering the continents, may be far more numerous in deeper waters than is generally realized. (b) Present geological conditions may well be responsible for artesian outflow on the relatively steep seaward face of certain sections of the continental shelves. (c) Past geological conditions were almost certainly far more favorable, in many if not in all areas, to such submarine artesian outflow, than are the conditions observed today. (d) Under such past conditions artesian outflows far down the seaward slopes of continental shelves may have been of enormous volume, and may have endured for enormously long periods of time.

Expulsion of Non-artesian Waters from Continental Shelves. Deposition of marine sediments to help form the continental shelves involved the inclusion of vast quantities of sea water. The amount of water thus incorporated has been variously estimated at from 25 to 50 percent of the volume of the sediment, except in the case of fine silts and clays which may contain much larger proportions. Seelheim[44] found experimentally that fine clay will settle in a stratified condition with a content of water equal to 79 percent of the volume of the sediment in the upper and 64 percent in the lower layers. Hedberg[45] states that "immediately after deposition muds may have porosities of 70 to 90 per cent, silts from 50 to 70 per cent, and sands from 30 to 50 per cent." In testing recent mud deposits along the sea coast, Shaw[46] found that the water content ranged "from 40 or 50 per cent to 90 per cent or more." Meinzer[47] quotes an unpublished communication by Shaw to the effect that much new Mississippi delta deposit has a porosity of 80 to 90 percent.

The quantities of sea water expelled during the consolidation of marine sediments must be enormous. King[48] estimates that the consolidation of 1,000 feet of compact limestone involves the expulsion of a quantity of water equal to not less than one-fourth and possibly as much as one-half of the present volume of the rock. The consolidation of silts and clays to form shales requires the expulsion of still greater quantities of water. Sorby[49] states that as a result of the squeezing out of water, shales and slates may occupy, in extreme cases, only one-ninth of the volume they possessed when deposited. Even sandstones may have had their storage capacity for water decreased by 20 percent or more. To the expulsion of water resulting from compaction must be added a further great quantity resulting from deposition of mineral matter to increase size of constituent grains and reduce pore space. Assuming a thickness of 36,000 feet of Palaeozoic sediments in the Appalachian region, and a present average pore space of 23 percent to be still occupied by water, King[50] estimates that the water displaced would give a sea 3,600 feet deep. He then continues: "The actual pore space of these rocks is now very much less than 23 per cent, so that there has certainly been a very large movement of water out of them since their deposition, a movement which must have been twice, if not thrice, the amount stated above, if such depths of sediments were formed."

If one compares the Atlantic continental shelf of North America with the Appalachian Palaeozoic sediments, it becomes evident that in respect both to original thickness of deposits and present degree of consolidation, conditions favor a much larger expulsion of water from the older beds. Nevertheless the total quantity of water expelled from the shelf sediments must have been enormous. King, Shaw, and others have recognized that waters expelled by compaction and by deposition of mineral matter will move upward or downward into the more pervious formations, and then laterally toward points of easiest exit. Thus there may have developed on the face of the continental shelf concentrated outflows of considerable volume at a limited number of points.

It must not be imagined that such outflows were merely a temporary source of submarine springs; that the flows were of vast proportions for a relatively short period when consolidation was in progress, and then ceased altogether when consolidation was completed. Both sedimentation and consolidation have been a continuing process, presumably occurring on one part or another of the Atlantic shelf of North America from Cretaceous or earlier times to the present. Doubtless there was more active outflow at one point during certain periods, at others during different periods. Thus we are not forced to choose between the conceptions of a vast total of water expelled with catastrophic effects during a single short period, or of the same outflow so evenly distributed over a great lapse of time as to give a constant but relatively insignificant development of submarine springs. When allowance is made for concentration of outflow at a limited number of points and for greater intensity of outflow at some times than at others, it seems reasonable to suppose that deep submarine springs due to consolidation of sediments may long have been an important phenomenon on the continental slope. Whatever the outflow due to this cause, it supplemented the artesian outflows earlier discussed.

Having demonstrated the reasonableness of the conception that submarine springs have played an important rôle in the history of continental shelves, particularly along their seaward borders, we turn next to consider the competence of spring sapping to produce canyons of great magnitude.

B. COMPETENCE OF SUBMARINE SPRINGS TO EXCAVATE CANYONS

The effectiveness of spring sapping in producing small valleys and even deep canyons on the surface of the earth has long been recognized. A few examples from the literature will suffice. Where rivers built large deltas into temporary glacial lakes, conditions often favored the development of large springs along the base of the frontal slopes. The exposed frontal slope of the Sheyenne River delta, built into former Lake Agassiz in North Dakota, shows the effect of sapping by such springs. According to Hall and Willard

the northeast front of the delta . . . . is intersected by several deep coulees which have been formed by the action of springs bursting out from the delta. These may fittingly be called "traveling springs," since they travel backward into the plateau as a result of the action of their own waters in removing the erodable materials out of which they emerge. . . . . The spring half a mile west of Leonard village has eroded a gorge 2 miles in length with a maximum depth of 70 feet. Other coulees in the vicinity formed in the same manner are half a mile to nearly 2 miles in length.[51]

Professor Henry S. Sharp recently directed my attention to the effects of much more extensive spring sapping observable in northwestern Florida (Figure 4). Here the "traveling springs" have cut back into the upland of the Coastal Plain by amounts varying from 2 or 3 miles up to 8 or 10 miles or more, giving gorges 50 to 100 feet deep. As there is little correlation between the heads of the gorges and the irregularly undulating surface topography, one can not attribute valley development to surface drainage, but must accept as valid Sellards's explanation that spring sapping is the chief factor in headward growth of the valleys.[52] Valley heads are often steep-sided amphitheaters, sometimes almost cirque-like in appearance, and are locally called "steep-heads." Springs issuing at the base of the amphitheatral steep-heads are sometimes so large as to be shown on topographic maps (see Holt topographic quadrangle, Florida).

A more impressive instance of canyons above sea level produced by spring sapping is found in the "alcoves" bordering the Snake River in its course through the lava plateaus of southern Idaho (see Frontispiece). In 1902 I. C. Russell[53] described these "box-headed canyons," varying in length up to two miles and having depths of several hundred feet. At their heads are springs eleven of which rank among the largest in the United States. Russell concluded that alcoves are the product of spring sapping, a conclusion fully confirmed by the later and more detailed studies of Stearns.[54]

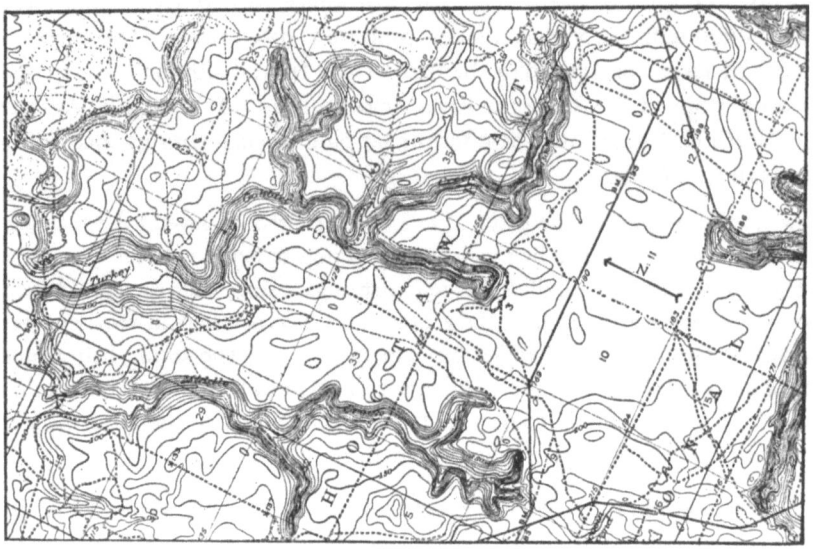

FIGURE 4: "Steep-head" valleys produced by spring sapping in the coastal plain of northwestern Florida. Reproduced from Holt quadrangle, U. S. Geological Survey.

In an earlier chapter of this volume attention was directed to Bryan's belief, reported by Stetson,[55] that canyons of the order of magnitude of submarine canyons on Georges Bank can be developed above sea level by groundwater sapping. Portions of Bright Angel Canyon were cited as examples. In commenting on Bryan's suggestion we then wrote:

Groundwater issuing as springs in valley heads is a normal accompaniment of the headward growth of valleys, and where the geological structure favors the undermining of higher beds, sapping is apt to occur. In any event the major point made by Bryan seems incontrovertible: notches cut many thousands of feet deep and a number of miles back into the face of a plateau scarp do not imply the former presence of large streams. Such notches are common features in the faces of mountain and plateau scarps the world over, and usually in circumstances which preclude the possibility of their erosion by streams of any considerable magnitude. . . . . The depth of the canyons depends on the height of the scarp, not on the size of the stream; and the length of the canyons depends on the length of time erosion has been in progress, not on the magnitude of the erosive agent.

The canyons of the Bright Angel region are 4,000 feet or more in depth, and a number of miles in length. They have been formed in a comparatively brief period of geologic time, and given more time, the canyons will undoubtedly become deeper and very much longer. Inasmuch as intermittent streams on the Colorado Plateau upland flow into the heads of the Bright Angel Canyons, we can not be sure that these deep notches are wholly the product of spring sapping. But I see no reason to question Bryan's belief that canyons of this and even greater magnitude could be produced by spring sapping alone. Hence, while spring sapping above sea level does not offer a satisfactory explanation for submarine canyons because of the enormous world-wide vertical oscillations of land or sea level required, it does help us to understand that, given favorable geological and other physical conditions, spring sapping is competent to produce canyons of any magnitude.

The crux of the matter lies in the question as to whether physical conditions under the ocean are such as to permit spring sapping on a large scale. That it does occur, and is a frequent cause of cable breaks, is believed by many students of submarine conditions. But it is one thing to have local undermining and slumping produced by suboceanic springs, and quite another to have long and deep canyons produced by their action. On land there is obvious opportunity for removal of the products of erosion and slumping. It is by no means obvious how such products can be disposed of under the sea, and the development of canyons thus proceed indefinitely. Nevertheless, analysis of the problem will demonstrate that canyon development by spring sapping must be recognized as a possibility under the sea as well as on land.

Rôle of Solution in Spring Sapping. When one speaks of stream erosion, one instinctively pictures a more or less vigorous current of water transporting a significant quantity of debris which is employed as an abrasive in cutting vertically and laterally into the country rock. One also thinks of debris from the rock walls slumping and creeping downward to the river, there to be further eroded perhaps, and in any case ultimately to be transported from the region in suspension and by traction. Such a picture seems hardly compatible with conditions believed to obtain in the ocean depths, where vigorous current action for any great distance along valley bottoms is difficult to imagine, and where deposition rather than erosion is believed to be the rule.

Even in a limestone region, surface canyons are usually pictured as primarily the result of mechanical erosion by streams armed with debris which serves as cutting tools. But, curiously enough, if the stream flows in an underground channel or tunnel, however great the dimensions of the latter, one does not hesitate to ascribe the formation of the tunnel primarily to solution. In many descriptions the fact that the subterranean stream carries debris and must abrade its channel is apparently overlooked, and the whole work of rock removal is explicitly or by implication assigned to solution.

Obviously there is no sufficient justification for such a mental attitude. It is more logical to keep clearly in mind the fact that both abrasion and solution play a rôle in the carving of every stream channel and valley, whether above ground or below; to recognize that *soluble* and *insoluble* are relative terms, and that the most "insoluble" rock is in fact partially dissolved by pure water as well as by water charged with CO_2 or other chemical substances; to recognize, further, that the proportions of abrasion and solution involved in the cutting of any given valley or canyon are difficult if not impossible to determine; and to pass on to the inevitable conclusion that, *given sufficient time*, the longest and deepest canyon in the most "insoluble" of rocks might be produced by solution alone.

As commonly employed in geology the term "insoluble" means *only slightly soluble under ordinary surface temperatures and pressures in the short period of geologic time during which the present surface features of our earth have been formed*. It is with this conception in mind that most igneous and metamorphic rocks, sands and sandstones (even when feldspathic), clays, and shales are usually regarded as "insoluble." Limestones, salt, and gypsum are classed as "soluble."

The student of submarine canyons can not expect to get a clear picture of all elements of his problem, if he approaches it with a mind under tyranny of conceptions adapted to another realm than that with which he is dealing. He has to do, not with ordinary surface temperatures and

Submarine Spring Sapping

pressures, and with short periods of geologic time, but with temperatures and pressures existing 5,000 to 10,000 feet below the surface of the earth's crust, and with periods of geologic time immensely long. If he can adjust his mind to these latter conditions, he may conclude that under them the relative importance of abrasion and solution as observed on the lands is reversed under the submerged continental shelves; and that along the deep seaward borders of the shelves solution may accomplish all that abrasion effects in canyons exposed to our inspection. And with the recognition of solution as a dominant factor, he may discover that most of the difficulties apparently confronting the hypothesis of canyon development by spring sapping are themselves dissolved.

To gain proper appreciation of the quantitative importance of solution during long periods of geologic time, let us review some facts brought out by studies of the solubility of minerals and rocks. Nearly a century ago the Rogers brothers[56] performed experiments to determine the extent to which some fifty varieties of rock-making minerals, rocks, and certain other materials such as glass, porcelain, anthracite, and wood were soluble in pure water and in water charged with CO_2. Their studies proved "most conclusively the solvent and decomposing power of pure and carbonated water upon all the important mineral aggregates, as well without as with alkaline ingredients." Finely powdered minerals and glasses, placed on pure filter paper and leached by carbonated water passing through them, commonly showed distinct evidence of partial solution in less than ten minutes.

In the case of *simple water,* the result is much feebler, and requires a longer time. But with nearly all the substances enumerated, it is entirely unequivocal, and with some of them quite intense. . . . . By the second method, that of prolonged digestion, we have actually made with carbonated water, and even with simple water, a partial analysis of a number of complex minerals. The specimens exposed to the CO_2 water for forty-eight hours, and to the simple water for one week, have in many instances furnished a sufficient amount of material to the liquid to admit of a quantitative examination.

In the case of hornblende, actinolite, epidote, chlorite, serpentine, and feldspar, from .4 to 1 percent of the whole mass was dissolved within the periods indicated. Müller[57] and Headden[58] later confirmed these results, Müller showing that various silicate minerals when treated with carbonic acid waters during seven weeks lost from .3 to more than 2 percent of their material.

In 1854 Bischof[59] in his *Elements of Chemical and Physical Geology* wrote: "Even the most compact rocks, which are scarcely, if at all, permeable to water, are decomposed in the interior of the earth by the carbonic water surrounding them, ... and the more rapidly when they are much rent." Bischof also showed that "meteoric water, while filtering through earths and rocks, dissolves quantities of substances by no means insignificant," and continued:

If this fact is duly considered, it will be evident what changes the meteoric water may effect in rocks, at one place taking up substances, at another depositing them by interchange for others, and thus after millions of years, producing results which excite our astonishment, although the means by which they are brought about are apparently so unimportant that they are mostly overlooked.

Referring to the Rogers brothers' demonstration that forty grains of finely powdered hornblende agitated in carbonated water for forty-eight hours lost 0.355 grains by solution, Bischof pointed out that were this treatment repeated 112 times with fresh carbonated water the whole forty grains would be dissolved in 224 days. Were the same quantity of hornblende in a solid mass instead of in powdered form so treated, complete solution would require somewhat more than six million years. But since hornblende is readily cleavable, frequently cracked and porous, permitting ready access of water, this period would under normal conditions be greatly reduced.[60] As Bischof further points out, the fact that one mineral is found in the crystal form (pseudomorph) of another supposedly insoluble is proof that the "insoluble" mineral was completely dissolved, even though the process may have required hundreds of thousands of years. He concludes: "All evidence, then, proves that those minerals whose constituents are not decomposed by the atmosphere, and suffer no alteration from this source, still are not capable of resisting the solvent action of water."

In a famous address delivered before the Liverpool Geological Society in 1876 and frequently quoted since, T. Mellard Reade[61] showed that the quantity of mineral matter annually removed in solution by the surface waters of England and Wales was sufficient to reduce the entire surface of those areas by one foot in approximately 13,000 years. Evenly distributed, the material so removed would amount to 143.5 tons per square mile of surface, "a figure surprisingly large." The average for the entire land surface of the globe he placed roughly at 100 tons per square mile, silica (7 tons) ranking fourth in the order of importance of dissolved constituents. With more and better quantitative data at his command, Clarke[62] has estimated the average solvent denudation of all the lands at one foot in 30,000 years.

After reviewing the work of various students of mineral and rock solubility Merrill[63] concludes: "These and similar tests by more recent workers show with apparent conclusiveness that all the ordinary rock-forming minerals,—silicates, oxides and carbonates—are appreciably soluble in the water of rainfalls and at ordinary temperatures." Merrill cites illustrations of marked solution effects observed by Hayes, Campbell, and Willis on quartz pebbles exposed in one case to drip from an overhanging cliff, in another to the water of a stream. In 1924 Clarke summed up the situation in these words:

> The evidence, both as found by experiment in the laboratory and by field observations, shows that practically all minerals, certainly all of the important ones, are attacked by water and carbonic acid. The pyroxenes and amphiboles yield most readily to waters, then follow the plagioclase feldspars, then orthoclase and the micas, with muscovite the most resistant of all. Even quartz is not quite insoluble, and the corrosion of quartz pebbles in conglomerates has been noted by several observers.[64]

Cushman[65] studied the cementing value of many varieties of powdered rock, including granite, gneiss, diabase, basalt, syenite, quartzite, chert, sandstone, slate, argillite, and arkose. In the course of his experiments he found that nearly all varieties of rock show distinct solution effects when their powders are ground for three hours with approximately 20 percent by weight of water.

Once the solubility of minerals and rocks is established, the competence of running water to create great canyons becomes merely a question of time. If the minerals are highly soluble, as in the case of limestone, the time required will be extremely short. Hence the ease of recognizing solution effects on limestone in areas but recently subjected to subaërial denudation. If the minerals are moderately soluble, as in the case of basic lavas, the time required will be much longer. Hence the tendency to think of basic lavas as insoluble, and the consequent long delay in recognizing the existence of "lapiés" or solution channels in the volcanic rocks of Hawaii,[66] and the solution origin of large canyons in the lava plateaus of Idaho.[67] If the minerals are but slightly soluble, as in the case of some acid lavas, gneisses, sandstones, and shales, the time required will be immensely long. Hence the almost complete failure to recognize the possibility of solution channels, ravines, and canyons in rocks of this type. Yet, *given time enough,* solution can achieve in the least soluble rocks effects produced in more soluble rocks in shorter periods of time.

With this fundamental fact in mind let us again examine those examples of valleys and canyons formed by spring sapping briefly described on an earlier page. In all but one of the cases cited the usual assumption, that solution has played a subordinate rôle in the fashioning of surface valleys, is not specifically called in question. But in the case of the impressive alcoves (Frontispiece) bordering the Snake River in Idaho, Stearns has presented evidence which seems to establish his conclusion that spring sapping by solution has been the dominant cause of canyon development. That spring sapping was responsible for the canyons was early recognized by Russell, who wrote:

> The origin of the side canyons which receive no surface streams, such as Little and Box canyons and the one in which the Blue Lakes are situated, is plainly due to the action of the great springs which come out at their heads. . . . . The great springs undermine the basalt by removing the soft material on which it rests. Thus blocks of the usually more or less vertically jointed rock break away and fall into the spring, and sooner or later sink into the soft bed beneath, as the emerging waters remove the silt and sand from beneath them. By this process vertical walls without talus slopes are produced. This process continuing, the cliff recedes, leaving a side cut or alcove in the wall of the main canyon, which becomes lengthened into a lateral canyon. The process would seem to be cumulative, as the farther the head of a side canyon receded the greater would be the tendency of the escaping waters to converge toward it.[68]

Russell believed that "the soft material" under the basalt, removal of which permitted sapping, consisted of beds of white silt and fine quartz sand, since he found such material in the Blue Lakes and other springs. He pointed out, however, that "no exposures of the material on which the basalt rests are known to me."

Stearns[69] found that the white sand in Blue Lakes springs was typical wind-blown sand from the Snake River flood plain; that the basement at Blue Lakes alcove was relatively impermeable basalt rather than sand; that andesite was the basement in other alcoves; that tunnels driven to intercept the spring waters show the latter moving through interstices in basalt where the latter rests on an impermeable basement; that exposures of alcoves where their basements are high above river level and dry fail to reveal the talus blocks postulated by Russell as having been engulfed in a bed of sand; while the absence of exposures of any such bed of sand or silt remains a weak point in Russell's hypothesis.

Ordinary erosion of the canyons is eliminated by Stearns. As he points out, the springs are clear and have low velocity at their points of issue; yet it is precisely here, where sapping is active, that talus is most scanty. Down-canyon from the springs the water in places flows swiftly, and is armed with quartz sand blown from the adjacent river valley; yet talus formed along side walls of the canyons is not removed as rapidly as at their heads.

Stearns concluded that the blocks of basalt, falling into the spring pools as sapping progresses, are chemically weathered and slowly dissolved. He concentrated attention on the amount of solution effected in the pools, rather than on solution during passage of the waters underground, since the competence of water to remove talus blocks shed into the pools was the chief point at issue. Analyses of a few samples of water from the springs and from adjacent wells failed to show any appreciably higher mineral content in the spring waters, a result which could have been predicted. But, as Stearns observes, "the volume of water discharged by these springs is so great that if the water dissolved only 5 parts per million of solids after it issued, solution would be competent to form the alcoves in the estimated time since they originated."

Stearns's observations do not necessarily eliminate the possibility that solution and disintegration of the basalt resulted in transport downstream of some finer disintegration products in suspension. This possibility is explicitly recognized in a later paper by the same author and two associates.[70] In any case Stearns's final conclusion, that the canyon-like alcoves could be produced by solution of the spring-sapped debris, seems fully justified.

According to the history of events set forth on page 448 and in Figure 8 of Stearns's first paper, the Blue Lakes springs developed after extrusion of the Sand Springs basalt, dated well along in the Pleistocene. If a canyon (see Frontispiece) two miles long and 300 feet deep at its head can be developed by solution in basalt during a part only of Pleistocene time, how large a canyon could be developed by solution in sedimentary beds during the immensely longer period in which outer portions of the continental shelves may have been subjected to spring sapping?

Submarine Spring Sapping

When seeking an answer to such a question many variables must be considered. Some sedimentary beds are far more soluble than basalt, others far less so. It is worthy of note that in the Atlantic Coastal Plain one of the most notable artesian horizons, the Tuscaloosa formation, located near the base of the Upper Cretaceous and consisting in considerable part of arkosic sands, is relatively so soluble and transmits water so freely that sinkhole topography is developed over its surface outcrops.[71] Some overlying beds are more soluble than the arkose, others less soluble. But so far as known most of the coastal plain–continental shelf deposits consist of fragmental material between the particles of which water can penetrate and facilitate the process of solution. Canyon development does not, however, depend upon complete solution of the rocks attacked. The finer products of disintegration may be removed to some distance in suspension, while, as will later appear, both fine deposits and coarser debris may repeatedly be discharged from canyons in the form of mudflows. There seems to be no sound reason for excluding the possibility that at the heads of submarine canyons, just as at the heads of alcoves on land, outflowing waters can, given sufficient time, dissolve, disintegrate, and remove debris dropped into the canyon heads as a result of spring sapping.

Certain factors among the variables to be considered appear peculiarly favorable to spring sapping of shelf borders with accompanying solution. Pressures at depths of 5,000 to 10,000 feet must expedite solution. Higher temperatures at these depths must greatly promote solvent action. Whether the mineral content of artesian and other outflowing waters will aid or retard solution depends on the character and quantity of such content. Joly[72] has demonstrated that sea water is from two to fourteen times as effective in dissolving rock-making minerals as is fresh water. Hence outflowing connate waters may be expected to be a relatively good solvent, while the walls of canyons exposed to sea water for long geologic periods must suffer solution and disintegration favorable to slumping of their steep sides. Davis[73] has recognized the rôle that "weathering" of rock by sea water may play, although he invoked submarine currents for removal of the weathered debris. In discussing "mock valleys" (submarine canyons) he wrote:

> Are we justified, for example, in rejecting a slow process of submarine erosion by which mock valleys could actually be excavated in the slopes of the sea floor? May not the rocks in certain parts of the submarine continental

Submarine Spring Sapping

slopes be slowly disintegrated by the high-pressure sea water? May not an outflowing sea-floor current . . . . be competent to remove the leached slime from rocks thus disintegrated? Would such action be more remarkable than the removal of silica from decomposed feldspar in rain-water solution during the production of laterite on subaerial peneplains, the degradation of which requires hundreds of thousands of years? If such submarine erosion . . . . is possible at all, might it not, in the course of a geological period, produce significant changes of sea-floor configuration?

A favorable factor of great importance is the circumstance that solution channels in the submerged shelf, whether minute or of grand dimensions, are always filled with water. Davis[74] has shown how important is this circumstance in the solution of limestone caverns below ground-water level. Where artesian or other waters are moving through the shelf deposits, all parts of the channel-ways, bottom, sides, and top, are constantly subjected to solution, instead of a mere fraction of the bottom as in land valleys or subterranean channels above groundwater level. If channel-ways in the submerged shelf become relatively large and the flow of water is relatively rapid, mechanical erosion may supplement solution, and the outflowing waters may then bear a load of sediment in suspension or even dragged along the channel floors. Deposition of this sediment beyond the spring exits would tend to clog the canyon heads; but, as later pointed out, mudflows of the saturated debris would tend to set limits to such accumulation.

If one could properly evaluate all the variable factors involved in solution and disintegration within the shelf, and spring sapping, solution, and disintegration along its deeply submerged border, the conclusion might well favor more active development of canyons at shelf margins than on land areas like that of the alcoves bordering the Snake River. But whether conditions are more favorable on land or under the sea, there seems to be no escape from the conclusion that, *given a sufficiently long lapse of geologic time*, canyons of maximum dimensions may be produced by spring sapping along the submerged border of continental shelves.

To appreciate this fact we must, it is true, fundamentally alter the attitude of mind appropriate to the more rapid types of geologic changes ordinarily witnessed on land. We must, as already noted, completely revise our definitions of soluble and insoluble rocks. We must remember that even the grandest canyons on the earth's surface have been cut in an exceedingly minute fraction of geologic time. We must think of canyons forming under the sea during periods sufficiently long for vast areas of the lands to be reduced to peneplanes.

It required much effort before geologists could accustom themselves to the conception of a slow denudation of the land, largely by solution in certain stages of the process, to a low-lying plane at or near sea level. Today that conception is widely accepted, and peneplanation is recognized as a fact to be dealt with in earth history. It took many years of study to train even the scientific mind to think in terms of geologic time sufficiently vast to account for the erosion accomplished on the lands, the deposition of sediments piled up in the sea, and the slow evolution of animal forms entombed in those sediments. Small wonder then if the mind hesitates to recognize in submarine canyons, thousands of feet deep and miles in length, the product of slow sapping and solution throughout a vast period of time. As King[75] said in his discussion of groundwater movements: "It is only when we consider these slow and long processes of nature in their quantitative relations that we can realize how large and how important they have been and still are."

Fragments of rock containing Cretaceous fossils have been dredged from the walls of submarine canyons near the outer edge of the North American Atlantic continental shelf. Whether parts of the outer shelf are still older remains to be discovered. In any case expulsion of sea water trapped in the sediments must have begun in Cretaceous time, and outflow of artesian waters may have begun as early. Spring sapping and solution may thus have been more or less continuous for the last sixty million years, possibly much longer, with alternating periods of increased and reduced activity.

In the absence of definite knowledge respecting the growth of the shelf during Cretaceous, Tertiary, and later times, we can not picture with any certainty the history of the sapping process. But we can at least consider possibilities. When spring sapping first started it was presumably active at countless points along the margin of the shelf. But just as on land countless small surface streams contending for drainage area soon give place to a relatively few most favored streams at widely separated intervals, so at the shelf border countless springs contending for subterranean drainage should soon produce by sapping a limited number of canyons more widely spaced. After the lapse of sufficient time there should be found: a few canyons cut very far back where conditions were specially favorable; more canyons of intermediate length which were successful for a considerable period, but ultimately lost drainage as the most successful canyons pushed farther back into the shelf; and a multitude of incipient canyons or ravines which were quickly outdistanced by their larger neighbors. The seaward face of the shelf would thus be "furrowed" with many parallel or nearly parallel ravines of variable depth; but also trenched at intervals by true canyons, some of moderate size, a few of major dimensions. The shorter ravines might branch moderately; but ordinarily the branches would not diverge widely from the general trend of the continental slope. The larger canyons could develop important and widely divergent branches in case subterranean drainage from either side favored, at least temporarily, a major development of lateral spring sapping.

Rôle of Mudflowing. If deposition of fine silt continued over the outer shelf and down its steeper seaward slope, results would vary according to the areas affected. The nearly level upper surface of the shelf would be built upward, so long as currents due to waves or other causes did not remove the deposits to other areas. Deposition on the "furrowed" seaward face of the shelf might tend to blanket and partially smooth out inequalities of that surface, since spring action would no longer be active in the inter-canyon segments of the face. But a new factor would now come into play, tending to preserve the original furrows, to deepen them, or in some circumstances to form new ones. Silt deposits contain a high percentage of water when first laid down, up to 90 percent of their volume according to authorities earlier cited. Such a sheet of relatively fluid mud may remain stable on a level or almost level surface. But, on the steeper seaward slope of the shelf, a flowing of the deposit must be expected after a limited amount of deposition, if we are to credit such studies of subaqueous movement as that by Hadding.[76] Ravines due to earlier spring sapping, or threads of swifter flow in the new deposit, would draw the mud drainage into lines roughly parallel to the direction of slope, thus perpetuating an old system of grooves or producing a new one.

As mud flowed to the bottom of the slope and spread out at its very low angle of repose, this lower and gentler slope would be built up until it became the seaward face of the shelf. But just so long as the seaward face was steeper than the upper surface of the shelf, and material was carried from the upper surface out over its edge to be deposited upon and thereby steepen the higher levels of the seaward face, just so long would the seaward face be subject to recurrent mudflowing, and so preserve its ravined or fluted character. The process would not be identical with that responsible for the parallel "canyons" artificially produced on a miniature shelf face by withdrawing "ocean" waters in the experiments of Stetson and Smith;[77] but the appearance of fluting produced in those experiments gives a rough conception of what might result from initial spring sapping or parallel mudflows on the face of a real shelf.

Conditions in submarine canyons during deposition are of vital interest in this connection. Would these canyons be filled with the accumulating sediment? This scarcely seems probable. Silt deposited on the canyon bottoms would there give rise to flows of the relatively fluid mud, just as on the seaward face of the shelf. As a number of writers have recognized, such flows would evacuate the canyons at intervals, possibly carrying with them any coarser material washed in from the upper surface of the shelf, discharged by suboceanic springs, or fallen from the steep canyon walls. It is even conceivable that the flowing mud, especially if armed with debris from the walls, might erode and deepen the bottoms of the canyons.

Such intermittent flows of mud, consequent upon periods of limited deposition, should not be confused with the conception of landslides or mudflows from previously filled canyons, discharged following earthquakes or other disturbance. Canyon filling by silt deposition could occur only in case the gradients of the canyon floors were extremely gentle. Even in such case the canyon filling would involve concomitant building up of the adjacent ocean floor. Otherwise the accumulation of relatively fluid mud in the canyon would be disgorged from the canyon mouth whenever increasing weight of the deposit overcame internal friction and initiated sliding or flowing.

Filling of canyons by coarser debris less subject to flowing does not appear likely, except where glaciers approached sufficiently close to have dropped their debris in or near the canyons; or where canyons head so close to shore as to receive coarse debris moved by littoral currents. Under normal conditions, spring sapping will not cut back the head of a canyon any faster than solution and disintegration permit removal of the sapped debris. Material discharged in the spring waters will be already in solution, or suspension, or subject to solution and disintegration when it collects about the orifice, unless earlier removed by mud-

Submarine Spring Sapping

flowing. In either case up-welling spring waters must tend to keep fluid the accumulating mud, and thus to favor canyon evacuation by recurrent mudflows.

Should the outflowing spring waters be lighter than sea water they will rise toward the surface, corroding the headwall of the canyon so long as they are in contact with it, otherwise diffusing in the sea water as they rise. But should they be heavier than sea water by virtue of their content of dissolved mineral matter or for other reasons, they may flow seaward along the canyon bottom with sufficient velocity to accelerate solution there and even to help keep the canyon clear of accumulating sediment.

While it thus seems within the bounds of reason to suppose that Nature has provided a mechanism for the carving of canyons under the sea, and for keeping them open once they are carved, it is not to be supposed that such mechanism works everywhere, or with equal perfection in those places where it does operate. We must be prepared to find that canyon development has been active in some places, sluggish in others, inoperative in still others.

C. FORM AND DISTRIBUTION OF CANYONS COMPATIBLE WITH SPRING-SAPPING HYPOTHESIS

If we examine the available information relating to submarine canyons cut in the margins of continental shelves, we find a significant degree of correlation between features actually known to exist and features expectable under the hypothesis of canyon development by the sapping action of artesian springs. This statement does not apply to certain long and narrow "deeps" in the ocean floor which have by some been called "submarine canyons"; nor to certain shallow trenches traversing the whole, or nearly the whole, breadth of the continental shelf, sometimes even entering the estuaries of great rivers. These forms presumably have distinct modes of origin, and may preferably be called by other names than submarine canyons, even when they appear to be connected with such canyons. There are doubtless other submarine features, more or less canyon-like in form, which deserve to be classified apart and studied as independent problems. Here we are concerned solely with that large group of true canyons cut in the margins of continental shelves in various parts of the world.

The distribution of such canyons is world-wide except for their *appar-*

ent absence in the Arctic and Antarctic regions. On the other hand, the distribution is locally erratic; for example, many canyons are found off the northeast coast of North America, none thus far off the southeast coast. Under the hypothesis here considered, canyon development through submarine sapping by artesian springs should in general be a world-wide phenomenon. But canyons should locally be poorly developed or absent where structural and stratigraphic conditions do not favor artesian flow, where the artesian head was not sufficiently high for a sufficiently long period of geologic time, or where glaciers, marine currents, or other agents have had opportunity to fill canyons formerly existing.

The occurrence of numerous closely spaced more or less nearly parallel minor canyons or ravines on the seaward face of the shelf, with canyons of moderate or major size interpolated at irregular intervals, as shown on a published section (preliminary) of Veatch's map of the North Atlantic shelf,[78] is in accord with expectations under the hypothesis of spring sapping. As we have already seen, numerous minor canyons should be produced in the early stages of spring sapping by waters expelled from consolidating sediments or by artesian waters, and should be preserved, intensified, or even redeveloped by recurrent mudflows where sedimentation subsequently takes place on the face of the shelf. A moderate number of these initial canyons should win enough subterranean drainage to develop into canyons some miles in length, and occasional specially favored examples into canyons of major dimensions. Observed facts, and expectations deduced from the hypothesis, are thus in accord, so far as distribution is concerned.

The close resemblance between submarine canyons and terrestrial river valleys is fully accounted for. Spring sapping on land produces valleys and canyons similar in form to those produced by ordinary stream erosion. Valleys and canyons of both origins show similar patterns as well as similar forms.

Sapping by springs readily accounts for the very steep walls reported from many submarine canyons, and for the occasional discovery of what appear to be landslide masses in some of them. Such accidents, and slumping due to subterranean solution, may interrupt normal canyon gradients, giving the local "deeps" observed in some places, the shallows and "steeps" noted in others, usually, if not always, on the basis of soundings so limited as to leave some doubt as to their reality. Mudflows and spring sapping have already been invoked to account for the re-

Conclusion

ported rupturing of cables on, or near the base of, the seaward face of the shelf. The statement that cable breaks unrelated to earthquake disturbances are most frequent off certain coasts following seasons of maximum rainfall does not seem wholly fanciful, since springs not vigorous at other times, and mudflows requiring only slight disturbance to "set them off," could be called into action when outflow of subterranean waters is most abundant. The finding of rock fragments on the ocean floor at the foot of continental slopes can readily be explained if mudflows debouching from the canyons carry debris which was either dropped from steep walls or washed in by marine currents or subterranean waters.

The hypothesis here under consideration avoids the difficulty of "too little time" involved in explanations assigning a comparatively recent date to the canyons, since it carries the beginning of spring sapping and mudflowing back into the Cretaceous, possibly even farther back. It also avoids the drastic assumptions of vast oscillations of land level or sea level for which there is no supporting evidence and against which many facts speak eloquently. It accounts for the fact that many canyons appear to be unrelated either to shallow surface channels on the shelf or to large rivers on the present mainland. But it recognizes that where land rivers are numerous many of the canyons must by chance appear more or less nearly opposite their mouths, and that where shallow trenches traverse the continental shelf some canyons may cut back into them.

CONCLUSION

The hypothesis of submarine canyon development by the sapping action of artesian springs should not be confused with those hypotheses, occasionally encountered in the literature, which invoke erosion by local and wholly mythical "subterranean rivers" discharging upon the ocean floor; erosion by surface streams supposed to disappear underground behind sand bars and to reappear at great depth farther out to sea; local prevention of deposition, or solution and erosion of soluble beds by waters upwelling along fault planes, as supposedly is the case in the famous "fosse de Cap-Breton." Rather it invokes normal and widespread conditions already known to exist over considerable areas of the earth's surface, but so far as the writer has ascertained not previously involved in the discussion of submarine canyons.

It should further be clearly understood that the present hypothesis is not offered as a fully established explanation of submarine canyons. It is offered rather as a working hypothesis which in the writer's opinion deserves critical study by all who are seeking to discover the origin of such canyons. So far as the writer can judge, the new hypothesis accords better with known facts than does any one of the hypotheses previously presented.

A decisive test of the new hypothesis is difficult to find. Exploration of hydrographic conditions in the bottoms of submarine canyons may be prosecuted with a view to discovering whether sea water there shows deficiencies or excesses of salinity, or abnormal temperatures, suggestive of outflowing subterranean waters. But it should be realized that while positive results must enormously strengthen the hypothesis, negative results will not weaken it, both because of the great difficulty of locating spring exits in the deep ocean, and because conditions today are admittedly far less favorable to artesian outflow than they were in earlier geologic times. Studies should be made where possibilities of present artesian outflow are best.

Experiments may be devised for provoking mudflows in canyon bottoms by discharging explosives, and recording the fact of flow by suitable instruments or by new soundings. But here again we face the difficulty that while positive results would go far toward establishing one phase of the hypothesis, negative results would prove nothing, since according to the hypothesis mudflowing takes place only under certain favorable conditions.

Whether or not the new hypothesis stands the test of critical analysis by other investigators, and eventually proves to be a reasonably satisfactory explanation for one of the most puzzling features on the surface of our earth, it is hoped that the review of various hypotheses attempted in this essay will prove of service to all students of the submarine canyon problem. For the primary purpose of this discussion is not to establish the validity of any particular hypothesis of canyon origin. It is rather to make available such a critique of what has thus far been accomplished in this field of enquiry, and to suggest such new possibilities of research, as will form a useful background for future investigations of a problem long unsolved.

Perhaps the ultimate solution will be found in some combination of certain of the hypotheses discussed on earlier pages. For the author to have made all possible combinations, and then to have analyzed the strong and weak points of each such combination, would greatly have extended a treatment already over-long. The chief need now is not for more hypotheses or combinations of hypotheses, but for more facts. These are being brought to light in increasing number every year. When we know more about submarine canyons the task of deciphering their history may prove less difficult.

Notes and References

1. M. J. de la Roche-Poncié, "Rapport sur la fosse et le havre de Cap-Breton (1860)." *Serv. Hydrog. de la Marine (France). Rech. Hydrog. sur le Régime des Côtes.* (Paris) Cahier 2 (1858-1863), pp. 62-74, 1877. See p. 68.
2. John Milne, "Sub-oceanic Changes." *Geog. Jour.*, Vol. 10, pp. 129-146, 1897. See pp. 143-144.
3. Henry Benest, "Submarine Gullies, River Outlets, and Fresh-Water Escapes beneath the Sea-Level." *Geog. Jour.*, Vol. 14, pp. 394-413, 1899. See pp. 395-397.
4. W. S. T. Smith, "The Submarine Valleys of the California Coast." *Science*, Vol. 15, pp. 670-672, 1902. See p. 672.
5. *Op. cit.*, p. 404.
6. Ch. Gorceix, "Sur la formation du 'Gouf de Cap-Breton.'" *Acad. Sci. Paris, Ct. Rend.*, Vol. 174, pp. 557-559, 1922.
7. *Loc. cit.*
8. Ch. Gorceix, "Le Gouf de Cap-Breton." *La Géographie*, Vol. 37, pp. 401-411, 1922.
9. P. E. Dubalen, "Eaux thermales des Landes et la fosse de Cap-Breton." *Actes de la Soc. Linn. de Bordeaux, Extr. des Proc.-Verb.*, Vol. 66, pp. 41-46, 1912.
10. Ch. Gorceix, "Sur la formation du 'Gouf de Cap-Breton.'" *Acad. Sci. Paris, Ct. Rend.*, Vol. 174, pp. 557-559, 1922.
———"Le Gouf de Cap-Breton." *La Géographie*, Vol. 37, pp. 401-411, 1922.
11. J.-B. Charcot, "Sur les températures à différentes profondeurs de la fosse du Cap-Breton." *Acad. Sci. Paris, Ct. Rend.*, Vol. 174, pp. 1246-1247, 1922.
12. H. C. Stetson and J. Fred Smith, "Behavior of Suspension Currents and Mud Slides on the Continental Slope." *Amer. Jour. Sci.*, Vol. 35, pp. 1-13, 1938. See p. 9.
13. Douglas Johnson, quoted by R. M. Field in "Structure of Continents and Ocean Basins." *Jour. Wash. Acad. Sci.*, Vol. 27, p. 189, 1937.
14. C. H. Hitchcock, "Fresh-Water Springs in the Ocean." *Pop. Sci. Mon.*, Vol. 67, pp. 673-683, 1905.
15. V. T. Stringfield, "Artesian Water in the Florida Peninsula." *U. S. Geol. Surv.*, Water-Supply Paper 773-C, pp. 115-195, 1936. See p. 152.
16. Élisée Reclus, *The Universal Geography*, Vol. 17. (London) 504 pp., [1876-1894] See pp. 362-363.
17. Hitchcock, *op. cit.*
18. Benest, *op. cit.*, p. 409.
19. Hugh Robert Mill, *The Realm of Nature.* (New York) 366 pp., 1894. See p. 240.
20. Konrad Keilhack, *Lehrbuch der Grundwasser- und Quellenkunde.* 3d ed. (Berlin) 575 pp. 1935. See pp. 211-213.
21. Karl Andrée, *Geologie des Meeresbodens.* Vol. 2. (Leipzig) 689 pp., 1920. See p. 237.
22. Milne, *op. cit.*, pp. 129-146 and (Part 2) pp. 259-284.

23. Benest, *op. cit.*
24. Andrée, *op. cit.*, p. 272.
25. Milne, *op. cit.*, pp. 143, 145.
26. Robert L. Jack, "Artesian Water in the Western Interior of Queensland." *Geol. Surv. Queensland*, Bull. 1, pp. 1-16, 1895. See p. 10.
27. V. T. Stringfield, "Ground Water Resources of Sarasota County." *Florida Geol. Surv.*, 23d-24th Ann. Rept., pp. 121-194, 1933. See pp. 162, 173-174.
28. P. N. Lynch, J. F. M. Geddings, and C. U. Shepard, Jr., "Report of Special Scientific Committee on Artesian Wells in and near the City of Charleston, S. C." *Year Book for 1881, City of Charleston, S. C.*, pp. 257-315, 1882. See pp. 271-272.
29. N. H. Darton, "Artesian Well Prospects in the Atlantic Coastal Plain Region." *U. S. Geol. Surv.*, Bull. 138, 232 pp., 1896. See p. 212.
30. Personal communication.
31. S. W. McCallie, "A Preliminary Report on the Artesian-Well System of Georgia." *Geol. Surv., Georgia*, Bull. 7, 214 pp., 1898. See p. 27.
32. L. W. Stephenson and J. O. Veatch, "Underground Waters of the Coastal Plain of Georgia." *U. S. Geol. Surv.*, Water-Supply Paper 341, 539 pp., 1915. See p. 448.
   Myron L. Fuller, "Contributions to the Hydrology of Eastern United States, 1903." *U. S. Geol. Surv.*, Water-Supply Paper 102, 522 pp., 1904. See p. 402.
   Waldemar Lindgren, *Mineral Deposits*. 4th ed. (New York) 930 pp., 1933. See pp. 49, 53.
33. Personal communication.
34. V. T. Stringfield, "Ground Water Resources of Sarasota County." *Florida Geol. Surv.*, 23d-24th Ann. Rept., pp. 121-194, 1933. See p. 174.
35. David G. Thompson, "Ground Water Supplies of the Atlantic City Region." *Rept. New Jersey Dept. Conserv. and Develop.*, Bull. 30, 138 pp., 1928. See pp. 114-115.
36. Douglas Johnson, *Stream Sculpture on the Atlantic Slope*. (New York) 142 pp., 1931.
37. N. H. Darton, "Artesian Waters in the Vicinity of the Black Hills, South Dakota." *U. S. Geol. Surv.*, Water-Supply Paper 428, 64 pp., 1918. See pp. 29-32.
38. Douglas Johnson, *The New England–Acadian Shoreline*. (New York) 608 pp., 1925. See pp. 264-304.
39. Joseph Barrell, "Criteria for the Recognition of Ancient Delta Deposits." *Bull. Geol. Soc. Am.*, Vol. 23, pp. 377-446, 1912. See p. 389.
40. Andrew C. Lawson, "The Isostasy of Large Deltas." *Bull. Geol. Soc. Am.*, Vol. 49, pp. 401-416, 1938. See p. 404.
41. A. W. Grabau, *Principles of Stratigraphy*. 2d ed. (New York) 1185 pp., 1924. See p. 612.
42. Richard Joel Russell, "Physiography of Lower Mississippi River Delta." *Louisiana Dept. of Conserv., Geol. Surv.*, Bull. 8, pp. 3-199, 1936. See p. 9.
43. Rollin D. Salisbury, *Physiography*. (New York) 770 pp., 1907. See p. 714.
44. F. Seelheim, "Methoden zur Bestimmung der Durchlässigkeit des Bodens." *Zeit. Analyt. Chemie*, Vol. 19, pp. 387-418, 1880.
45. Hollis D. Hedberg, "The Effect of Gravitational Compaction on the Structure of Sedimentary Rocks." *Amer. Assoc. Petrol. Geol.*, Bull. 10, pp. 1035-1072, 1926. See p. 1042.
46. E. W. Shaw, "The Rôle and Fate of the Connate Water in Oil and Gas Sands." (Continued discussion of the paper of Roswell H. Johnson.) *Amer. Inst. Min. Eng.*, Bull. 103, pp. 1449-1459, 1915. See pp. 1450-1452.
47. O. E. Meinzer, "The Occurrence of Ground Water in the United States." *U. S. Geol. Surv.*, Water-Supply Paper 489, 321 pp., 1923. See p. 8.
48. Franklin H. King, "Principles and Conditions of the Movements of Ground Water." *U. S. Geol. Surv.*, 19th Ann. Rept., Part 2, pp. 59-294, 1899. See p. 84.

Notes and References

49. H. C. Sorby, "On the Application of Quantitative Methods to the Study of the Structure and History of Rocks." *Geol. Soc. London, Quart. Jour.*, Vol. 64, pp. 171-233, 1908. See p. 214.
50. King, *op. cit.*, pp. 80, 81.
51. Charles M. Hall and Daniel E. Willard, "Description of Casselton and Fargo Quadrangles." *U. S. Geol. Surv.*, Folio 117, 7 pp., 1905. See p. 3.
52. E. H. Sellards and H. Gunter, "Geology between the Apalachicola and Ocklocknee Rivers in Florida." *Florida St. Geol. Surv.*, 10th and 11th Ann. Repts., 130 pp., 1918. See p. 27.
53. I. C. Russell, "Geology and Water Resources of the Snake River Plains of Idaho." *U. S. Geol. Surv.*, Bull. 199, 192 pp., 1902. See pp. 26-28, 127-130, 162-168.
54. Harold T. Stearns, "Origin of the Large Springs and Their Alcoves along the Snake River in Southern Idaho." *Jour. Geol.*, Vol. 44, pp. 429-450, 1936.
55. Henry C. Stetson, "Geology and Paleontology of the Georges Bank Canyons." Part 1. Geology. *Bull. Geol. Soc. Am.*, Vol. 47, pp. 339-366, 1936. See p. 353.
56. W. B. Rogers and R. E. Rogers, "On the Decomposition and Partial Solution of Minerals, Rocks, etc., by Pure Water, and Water Charged with Carbonic Acid." *Amer. Jour. Sci.*, Vol. 5, pp. 401-405, 1848.
57. Richard Müller, "Untersuchungen über die Einwirkung des kohlensäurehaltigen Wassers auf einige Mineralien und Gesteine." *Tschermaks Min. Mitt.*, pp. 25-48, 1877. See p. 25.
58. W. P. Headden, "Significance of Silicic Acid in Waters of Mountain Streams." *Amer. Jour. Sci.*, Vol. 16, pp. 169-184, 1903. See p. 181.
59. Gustav Bischof, *Elements of Chemical and Physical Geology.* Vol. 1. (London) 455 pp., 1854. See pp. 56-62.
60. *Ibid.*, p. 62.
61. T. Mellard Reade, "President's Address." *Liverpool Geol. Soc., Proc.*, Vol. 3, Part 3, pp. 211-235, 1877. See pp. 220, 224, 228-229.
62. Frank Wigglesworth Clarke, "A Preliminary Study of Chemical Denudation." *Smithsonian Misc. Coll.*, Vol. 56, No. 5, 19 pp., 1912.
63. George P. Merrill, *A Treatise on Rocks, Rock-Weathering, and Soils.* Rev. ed. (New York) 400 pp., 1906. See pp. 170, 172-173.
64. Frank Wigglesworth Clarke, "The Data of Geochemistry." 5th ed., *U. S. Geol. Surv.*, Bull. 770, 841 pp., 1924. See p. 484.
65. Allerton S. Cushman, "The Effect of Water on Rock Powders." *U. S. Dept. Agri., Bur. Chem.*, Bull. 92, 24 pp., 1905. See pp. 6-7.
66. Harold S. Palmer, "Lapiés in Hawaiian Basalts." *Geog. Rev.*, Vol. 17, pp. 627-631, 1927.
67. Harold T. Stearns, "Origin of the Large Springs and Their Alcoves along the Snake River in Southern Idaho." *Jour. Geol.*, Vol. 44, pp. 429-450, 1936.
68. I. C. Russell, "Geology and Water Resources of the Snake River Plains of Idaho." *U. S. Geol. Surv.*, Bull. 199, 192 pp., 1902. See pp. 127-130.
69. Stearns, *op. cit.*, pp. 445-447.
70. H. T. Stearns, Lynn Crandall, and W. G. Steward, "Geology and Groundwater Resources of the Snake River Plain in Southeastern Idaho." *U. S. Geol. Surv.*, Water-Supply Paper 774, 268 pp., 1938. See p. 146.
71. Laurence L. Smith, "Solution Depressions in Sandy Sediments of the Coastal Plain in South Carolina." *Jour. Geol.*, Vol. 39, pp. 641-652, 1931.
72. J. Joly, "Some Experiments on Denudation by Solution in Fresh and Salt Water." *Royal Irish Acad., Proc.*, Vol. 24, pp. 21-33, 1902. See p. 30.
73. W. M. Davis, "Submarine Mock Valleys." *Geog. Rev.*, Vol. 24, pp. 297-308, 1934. See p. 302.
74. W. M. Davis, "Origin of Limestone Caverns." *Bull. Geol. Soc. Am.*, Vol. 41, pp. 475-628, 1930. See p. 555.

75. Franklin H. King, "Principles and Conditions of the Movements of Ground Water." *U. S. Geol. Surv.*, 19th Ann. Rept., Part 2, pp. 59-294, 1899. See p. 83.

76. Assar Hadding, "On Subaqueous Slides." *Geol. Fören. Stockholm, Förh.*, Vol. 53, pp. 377-393, 1931.

77. H. C. Stetson and J. Fred Smith, "Behavior of Suspension Currents and Mud Slides on the Continental Slope." *Amer. Jour. Sci.*, Vol. 35, pp. 1-13, 1938. See pp. 10, 11.

78. A. C. Veatch, "Recent Advances in Marine Surveying: Work of the United States Coast and Geodetic Survey." *Geog. Rev.*, Vol. 27, pp. 625-629, 1937. See p. 626.

Résumé

Les bords extérieurs des plates-formes continentales submergées sont tranchés par de grands ravins dits "cañons sous-marins." Quelques-unes de ces fosses ont une profondeur de 2,500 mètres ou plus, une largeur de 3 à 6 kilomètres, et une longueur de 10 à 30 kilomètres, jusqu'à 50 à 100 kilomètres pour des exemples exceptionnels.

L'origine de ces formes remarquables est une question sur laquelle l'accord est loin d'être réalisé parmi les géologues et les géographes. L'auteur se propose d'examiner les hypothèses variées émises sur ce sujet, et de soumettre une nouvelle hypothèse qui lui semble posséder des avantages dignes d'être notés.

Les hypothèses d'une origine tectonique (dépressions formées par plissements ou par failles, fossés tectoniques) manquent d'expliquer d'une façon satisfaisante les cours sinueux des cañons, le profil transversal en V, le système arborescent des vallées tributaires, et d'autres formes ordinairement caractéristiques d'érosion fluviale, ainsi que l'absence de déformation importante de la plate-forme continentale au voisinage des cañons.

L'hypothèse la plus répandue prétend que ces formes sont des gorges coupées par des cours d'eau dans une période assez récente, quand les continents étaient beaucoup plus élevés, ou la mer beaucoup plus basse, qu'aujourd'hui. Cette hypothèse demande des oscillations du niveau de la terre ou de la mer d'une ampleur incroyable, et desquelles ni les formes côtières ni la surface de la plate-forme continentale ne donnent aucune indication.

On a supposé que ces ravins sous-marins peuvent être d'un âge géologique ancien, même plus ancien que la plate-forme continentale; qu'ils ont été submergés il y a longtemps; qu'ensuite ils ont été comblés par les sédiments marins; et enfin que ces sédiments ont récemment évacué les cañons par glissement sous forme d'éboulements sous-marins. Cette explication assigne à la plate-forme continentale une origine bien douteuse. Outre cela, on doit douter que les sédiments puissent être assez solides pour rester longtemps dans les gorges, même à leurs embouchures, et en même temps être assez fluides pour glisser dans les profondeurs de l'océan et évacuer les gorges sur des distances de quelques dizaines ou quelques vingtaines de kilomètres. On ne trouve pas, en face des gorges, les accumulations qu'on devrait attendre en vertu de cette hypothèse. Les évidences authentiques des éboulements au voisinage des cañons sous-marins manquent.

Les eaux souterraines peuvent saper les escarpements avant leur submersion, en donnant des ravins analogues à ceux qu'on trouve au bord du Grand Cañon du Colorado. Cette hypothèse rencontre les mêmes difficultés que celle de gorges fluviales récentes, discutées ci-dessus.

L'idée que les cañons submergés peuvent être les vides laissés par de simples éboulements sous-marins tombés de la face du grand escarpement continental n'est guère soutenue par personne.

De nombreux observateurs ont cherché à expliquer les cañons submergés en supposant qu'ils ont été creusés par des courants sous-marins. On examine alors les courants à marée, les courants hydrauliques produits en profondeur quand le vent pousse les eaux superficielles de la mer dans une baie, les courants produits par des différences de densité dues aux différences de température ou de salinité des eaux. On arrive à la conclusion qu'aucun de ces types de courants n'a la force de creuser de grands ravins sous-marins.

On examine plus à fond les courants produits par des différences de densité causées par des sédiments tenus en suspension. Tout récemment Daly a proposé que pendant les orages violents de la période glaciaire les vagues, se brisant sur la partie extérieure de la plate-forme continentale, plus exposée à cause d'un niveau de la mer plus bas qu'aujourd'hui, ont agité des sédiments fins. Ceux-ci, chargeant les eaux marines et les rendant plus lourdes, ont créé des courants d'eau trouble ("turbidity currents") qui en descendant le talus continental ont creusé les cañons sous-marins. La conception n'était pas nouvelle, et l'on trace ici son développement pendant un demi-siècle, depuis Forel jusqu'à présent. Forel, qui a conçu l'idée que les ravins sous-lacustres trouvés par Hörnlimann aux embouchures du Rhin et du Rhône dans les lacs Constance et Léman sont dus aux courants froids chargés d'alluvion et des matières en solution, a lui-même rejeté l'hypothèse que les cañons sous-marins ont été formés d'une manière semblable. Davis est le premier à invoquer des courants d'eaux troubles produits par les vagues et agissant pendant le niveau changeant de la mer de la période glaciaire, pour expliquer en partie les cañons sous-marins du côte de la Californie.

Les considérations favorables à l'hypothèse que les cañons sous-marins ont été creusés par des courants d'eau trouble, déjà exposés par Daly, sont brièvement résumés. Ensuite on démontre que les arguments présentés par cette autorité sont moins forts qu'ils ne paraissent pas. Au lieu d'appuyer l'hypothèse que les ravins sous-lacustres du lac Constance et du lac Léman ont été creusés par des courants d'eau d'une densité supérieure à cause de leur charge de sédiments, von Salis a attribué la différence de densité à la différence de température des deux eaux; tandis que Forel a invoqué une combinaison favorable des différences de température, de charge de matières en solution, et de matières en suspension. Forel a aussi considéré les ravins comme des

Résumé 115

produits d'une accumulation inégale de dépôts plutôt que comme des produits d'érosion. Il a rejeté explicitement l'idée que les cañons sous-marins de l'océan ont été formés d'une manière analogue aux ravins sous-lacustres du lac Constance et du lac Léman.

Les ravins sous-lacustres sont des phénomènes assez rares, tandis que les fleuves d'eau trouble débouchant dans des lacs sont nombreux. Évidemment le développement de ces ravins dépend de conditions bien particulières. Les cañons sous-marins n'ont pas les digues latérales qui caractérisent les ravins sous-lacustres, ni la plus grande profondeur (au-dessous de la surface adjacente) près de leur entrée. On est forcé de conclure, avec Forel, que l'hypothèse satisfaisante pour expliquer les ravins sous-lacustres ne suffit pas pour expliquer les grands cañons sous-marins.

Daly a cité les forts courants du détroit de Gibraltar pour renforcer l'hypothèse que des courants rapides peuvent résulter de faibles différences de densité. On montre que les courants du détroit résultent d'une combinaison de conditions favorables, et que la vélocité d'un courant dans un passage étroit entre deux continents n'est pas comparable à la vélocité des courants sur la face de la plate-forme continentale en pleine mer.

L'hypothèse que les grands cañons sous-marins ont été creusés par des courants de fond produits par la densité supérieure des eaux chargées de sédiment, exige qu'une grande quantité de sédiment reste en suspension dans l'eau de la mer pour des périodes assez longues. Mais suivant l'opinion la plus répandue des sédiments sont précipités avec une rapidité remarquable dans les eaux salines. Contre cette opinion Daly a cité les observations de Vernon-Harcourt et de Wheeler à l'effet contraire.

Pour résoudre cette question il faut (1) résumer quelques expériences classiques sur l'influence des sels différents en solution sur la rapidité de précipitation de sédiment; et aussi (2) faire attention à la question particulière examinée par Vernon-Harcourt et par Wheeler, et à la nature précise de leurs expériences. L'examen du premier point démontre que les recherches de plusieurs savants ont bien établi le fait que dans des eaux salines, comme celles de l'océan, les sédiments fins en suspension sont précipités assez vite,—beaucoup plus vite que dans des eaux douces. L'examen du second point démontre que Vernon-Harcourt et Wheeler ont tous les deux étudié une question bien différente: c'est-à-dire, si les matières des barres et des deltas aux bouches des rivières sont précipitées à cause des sels dans l'eau de la mer, ou plutôt à cause de l'arrêt des courants des rivières. Ils ont conclu, avec raison, que l'influence des sels dans la précipitation des matières des barres et des deltas est peu importante. L'objet de leurs études, et les méthodes employées dans leurs expériences, sont de tel caractère que leurs résultats, bien interprétés, n'offrent aucune réfutation des conclusions émises par des investigateurs de la période précédente.

Puisque les sédiments fins sont précipités rapidement dans l'eau saline, et puisque ces sédiments ne sont pas facilement pris en suspension dans la présence de sels solubles, il est difficile d'admettre que de tels sédiments peuvent charger l'eau de la mer en quantité suffisante, et pour une période de temps suffisamment longue, pour produire dans l'océan des courants d'eau trouble (turbidity currents) d'une force capable de creuser des cañons vraiment gigantesques.

Le degré de consolidation des sédiments qui composent les plates-formes continentales dans lesquelles sont creusés les cañons sous-marins est une question capitale pour l'étudiant qui examine l'origine de ces cañons. Si les plates-formes consistent de sédiments fins, pas encore consolidés et bien chargés d'eau, prêts à couler sur la moindre excitation, on peut concevoir que même des courants très faibles peuvent originer des mouvements importants. Mais si les dépôts qui constituent les plates-formes sont plus solides, et consistent au moins en partie de roche dure, l'érosion des grands cañons par des courants faibles dans une période de temps assez courte, paraîtra peu probable. On examine alors les faits connus jusqu'à présent sur ce point, et constate que les dépôts des plates-formes (sauf ceux sur la surface d'un âge tout récent) sont assez bien consolidés, avec quelques couches de grés et d'autres roches dures. Dans ces circonstances on hésite d'accepter une hypothèse qui exige le creusement des cañons profonds par des courants d'eau trouble (turbidity currents) qui doit avoir, comme nous l'avons vu, une force relativement faible.

L'eau de la mer, dans la zone d'agitation par les vagues, est moins saline et plus chaude que les eaux dans les profondeurs de l'océan. Ainsi les conditions sont tout à fait différentes de celles dans les lacs Constance et Léman, où l'eau des rivières s'enfonce rapidement sous les eaux des lacs. Dans l'océan les conditions de salinité et de température s'opposent à la naissance des courants d'eau trouble. Il est vrai que les rivières apportent à la mer une vaste quantité de sédiment en suspension, et Daly a cherché à trouver dans ce fait un autre moyen de créer des courants d'eau trouble sur le fond. Mais les témoins cités par beaucoup d'observateurs démontrent que l'eau trouble des rivières s'étale sur la surface de l'eau saline et plus dense de la mer sur des distances prodigieuses, et que l'eau marine reste assez claire sous l'eau fluviale. On doit douter, alors, si les eaux marines dans la zone d'agitation des vagues, mêlées avec les eaux fraîches venant de la terre, et moins salines et plus chaudes que les eaux plus profondes, peuvent descendre effectivement dans l'océan à cause d'une charge problématique de sédiment; et on doit douter également si les eaux marines peuvent devenir effectivement chargées de sédiment se précipitant d'une couche supérieure de l'eau fluviale trouble.

Résumé

Enfin, on considère plusieurs aspects généraux de cette question, et note d'autres difficultés qui s'opposent à l'hypothèse que les cañons sous-marins ont été creusés par des courants développant sur le fond à cause d'une différence de densité due à la présence de sédiment dans l'eau. On conclut que quoique cette hypothèse ne devrait pas être rejeté, mais au contraire mérite des études encore plus pénétrantes, elle est opposée par des considérations vraiment formidables. Par conséquent, on doit pousser l'enquête plus loin, et descendre dans les régions souterraines, et demander si là, au lieu de dans les royaumes subaëriens ou sous-marins, on peut trouver une solution de ce problème difficile.

L'hypothèse que les cañons ont été creusés par des fleuves souterrains débouchant sur le fond de l'océan a été émise il y a quatre-vingts ans. On doit admettre que de tels fleuves existent, plutôt dans les régions karstiques, et que quelques-uns peuvent déboucher dans la mer. Mais comme explication générale des cañons sous-marins cette hypothèse rencontre des difficultés évidentes, et n'est guère supportée par personne aujourd'hui.

Étroitement liée à l'hypothèse précédente est celle qui suppose qu'un cañon peut être produit par l'effondrement d'une caverne creusée par un fleuve souterrain. Cette interprétation a été invoquée seulement pour quelques cañons particuliers, et n'a jamais été acceptée, autant que nous sachions, comme hypothèse d'une application générale.

On peut supposer que des eaux souterraines montant le long d'une faille peuvent dissoudre des couches solubles en donnant des dépressions linéaires. Mais il ne paraît pas raisonnable d'attribuer à cette origine la plupart des cañons, étant donné leur forme en courbe et ramifiante, et leur relation systématique aux bords de la plate-forme continentale.

Gorceix a voulu expliquer la fosse de Cap-Breton par l'infiltration des eaux de mer dans une série de cassures coupant des couches d'argile, de sel et de gypse. Les eaux marines, en dissolvant le sel et le gypse et délayant l'argile, aurait provoqué une série d'éboulements en chapelet. Mais Gorceix a offert cette explication seulement pour les faits observés dans la fosse de Cap-Breton, "en dehors de toute théorie générale."

Dubalen cherchait à résoudre le mystère de cette même fosse en imaginant des eaux chaudes jaillissant d'une faille le long de la tranchée. Ces eaux chaudes "font naître de puissants courants de densité et de pression hydrostatique; ces courants suffisent à chasser les sables que les courants marins superficiels poussent à chaque instant dans la fosse." Cette explication n'a pas été acceptée comme hypothèse satisfaisante pour d'autres cañons.

Finalement, nous présentons une hypothèse qui autant que nous sachions n'a pas été considérée jusqu'à présent. Suivant cette hypothèse les cañons ont été produits par l'action des eaux, surtout artésiennes, qui pendant de longues périodes de temps ont jailli comme sources sous-marines, et qui ont sapé les bords de la plate-forme continentale en donnant les grandes tranchées. Quoique cette explication puisse paraître incroyable à premier coup, elle mérite d'être examiné, surtout quand toute autre hypothèse rencontre des difficultés apparemment insurmontables.

En examinant cette hypothèse il faut déterminer: (a) s'il y a une possibilité raisonnable que des sources sous-marines importantes puissent exister sur les bords extérieurs de la plate-forme continentale, à de grandes profondeurs; (b) si ces sources sont capables de former des cañons de grandes dimensions; et (c) si le caractère et la distribution des cañons sont compatibles avec cette méthode d'origine.

Il y a une abondance de sources sous-marines tout près des côtes, et parmi celles-ci quelques-unes de grandes proportions. Mais dans la nature des choses les sources à grandes profondeurs sont bien difficiles à découvrir. Néanmoins, l'étude de cette question conduit à la conclusion que des sources importantes doivent exister sur la pente extérieure de la plate-forme continentale, même à des niveaux très profonds. Beaucoup de géologues ont exprimé cette idée. L'examen des conditions artésiennes dans les plaines côtières et dans les plates-formes continentales démontre qu'au temps actuel des eaux artésiennes peuvent surgir à de grandes profondeurs. Les conditions favorables à cette action sont discutées au long. Beaucoup plus favorables auraient été les conditions dans le passé géologique, quand les couches perméables auraient dû avoir leurs affleurements aux hautes altitudes, et quand les eaux artésiennes auraient pu avoir issue à cinq ou dix milliers de pieds au-dessous du niveau de la mer, pour de grandes périodes de temps et sous une pression énorme. Les objections à cette conception sont examinées et réfutées.

Ensuite on examine les eaux non-artésiennes qui sont expulsées des sédiments en voie d'être consolidés. On trouve que la quantité des eaux ainsi expulsée était énorme, et que l'émission de ces eaux aurait pu continuer pendant de longues périodes de temps. Ces eaux ont supplémenté les eaux artésiennes en donnant des sources sous-marines qui ont dû jouer un rôle important dans l'histoire des plates-formes continentales.

On démontre alors que des sources ont excavé, par un processus de sapement (sapement = sapping), des vallées subaériennes assez importantes. La question est de savoir si ce processus peut opérer sous l'océan; et la difficulté capitale c'est de disposer des quantités énormes de débris excavés pour produire les grands cañons sous-marins. Cette difficulté disparaîtrait si le transport de ce débris était pour la plupart en solution.

Les expériences de plusieurs savants sont citées pour prouver que les miné-

Résumé

raux qui constituent les roches, aussi bien que les roches elles-mêmes, peuvent être dissous par l'eau pure, et plus rapidement par les eaux chargées de l'acide carbonique, de sel, et cetera. Par conséquent, on ne doit pas hésiter à conclure que le plus grand cañon sous-marin peut être excavé par sapement avec solution *à condition que la période de temps soit assez longue.*

Cette conclusion est renforcée par l'histoire du cañon appelé Blue Lakes alcove (Frontispiece) excavé en basalte par sapement des sources avec solution du débris, suivant les recherches de Stearns. Ce cañon imposant était produit dans un espace de temps géologique très petit, c'est-à-dire pendant la dernière partie du Pléistocène. Mais l'excavation des grands cañons sous-marins a pu commencer dans le Crétacé, peut-être même dans le Jurassique en quelques cas; et elle a pu continuer pendant soixante millions d'années ou plus. Aussi, la capacité des eaux artésiennes de dissoudre les roches est augmentée par la température et la pression qui existent aux grandes profondeurs, et peut-être par leur contenu de matières minérales.

Cependant, pour expliquer les grands cañons sous-marins il n'est pas nécessaire de rester exclusivement dans la dépendance de la dissolution. On peut invoquer l'abrasion dans les passages souterrains et tout près des sorties des eaux; le transport en suspension des matières très fines; et l'évacuation fréquente des sédiments déposés en face des sorties, par des éboulements de boue presque liquides qui peuvent charrier de chaque cañon les sédiments accumulés et quelque débris grossier tombé des parois raides ou déchargé par les eaux artésiennes. Ainsi, nous avons un méchanisme efficace pour excaver les cañons sous-marins, et pour les garder vides, pendant de vastes périodes de temps géologique.

La distribution des cañons sous-marins et leurs formes sont compatibles avec l'hypothèse qu'ils résultent de l'action des sources sous-marines qui ont sapé le bord extérieur de la plate-forme continentale. Les nombreux ravins qui sillonnent la face relativement raide (mais réellement assez douce) de la plate-forme peuvent être les résultats de sapement naissant, ou de coulées de boue constamment descendant la pente vers les profondeurs océaniques pendant toute la période de déposition des sédiments. Là où les sources ont sapé la plate-forme pendant une assez longue période, nous trouvons des cañons de dimensions moyennes.

C'est seulement quand une source a gagné une grande proportion de drainage souterrain qu'on trouve un cañon de la plus grande ampleur. Des sources latérales donnent naissance à des cañons en branche. Les éboulements de boue expliquent les ruptures des cables sur la face de la plate-forme et près du pied de la pente; aussi le fait supposé que dans certaines régions ces ruptures sont les plus fréquentes pendant la saison des grandes pluies.

Dans le texte on discute en quelque détail d'autres implications de la conception ici soumise à l'examen des géologues et géographes, pas comme une théorie bien établie, mais comme une hypothèse de recherche qui mérite, dans l'opinion de l'auteur, la considération sérieuse de tous ceux qui s'intéressent à la question de l'origine des grands cañons sous-marins.

Index

Abrasion: rôle in carving of stream channels, 94, 119
Acids: precipitating effects, 48
Adour River, 69
Africa: cable breaks, 74
Alcoves in lava plateaus, 71, 91, 98, 101
Alkalies: precipitating effects, 48
Allen, H. S., 51
Alpine rivers entering lakes, 42
Amazon River, 51, 56
Andrée, Karl, 74
Appalachian region, 82, 83, 84 (fig.), 90
Arkosic sands: solution of, 100
Artesian conditions, 72, 73, 75, 81; in continental shelves, 71, 74-81; in geologic past, 82-85; outcrop of horizons, 86-88; summary, 88-89, 118
Artesian wells: mineral content, temperatures, 78; low head, 80-81
Astronomical causes of sea level changes, 9
Atlantic City: artesian wells, 80
Atlantic Coast: continental shelf, 10, 25, 90, 102; location of canyons, 106; Veatch's map of shelf, 106
Atlantic Coastal Plain: artesian conditions, 81-85, 100; formations, 82, °3, 85, 87; fig. 84
Atlantic Ocean and Mediterranean Sea: difference in waters, 44 ff.
Australia: submarine springs, 73

Baltic Sea, 24n; bottom currents, 46
Bar and delta formation, 47-51 passim
Barrell, Joseph, 86
Barus, Carl, 48
Basalt, 98; canyon developed by solution in, 99, 119
Basic lavas, solubility of, 97
Bataillière, the, 43
Benest, Henry, 66, 67, 73, 74
Berkey, C. P., vi
Biscay, Bay of, 4, 30; see also Cap-Breton
Bischof, Gustav, quoted, 96
Blue Lakes: springs, 98, 99, 119
Bone, Captain, 13
Bottom currents, 24, 27, 36 ff., 42, 44, 46
Boulder Dam, 59, 60
Box-headed canyons, 92
Brahmaputra River, 56
Brech, the, 43

Brewer, William H., 48
Bright Angel Canyon, 92-93
Brown, J. S., vi, 79
Brunswick, Mo.: mineral well, 78
Bryan, Kirk, 18
Bullard, E. C., 85

Cable breaks, 12, 74, 87, 93, 107, 119
Cadız, Gulf of, 45, 46
California coast, 16, 55, 67; submarine canyons off, 31, 114
Campbell, M. R., 97
Canyons, maps, 10, 18, 50; competence of submarine springs to excavate, 91-105
Cap-Breton, fosse de, 4, 66-70 passim, 107, 117
Carbonated water, 94, 95, 119
Carolina bays, v
Caverns, subterranean: foundering of, 67, 117
Ceylon coast: submarine springs, 73
Charcot, J.-B., 70
Charleston, S. C.: artesian wells, 73, 77, 78
Clarke, F. W., 97
Clay, precipitation, 48; in canyon walls, 52 ff.; landslips in, 53; on seaward face of shelf, 87; water content, water expelled during consolidation, 89
CO_2: water charged with, 94, 95
Coastal plains: structure, 87, 91; see also Atlantic Coastal Plain
Collet, Leon W., 30, 42, 44
Columbia River: submarine valley, 14
Columbia University, vi
Compaction: expulsion of water resulting from, 89 f.
Congo River, 56; submarine canyon, 4, 30
Constance, Lake: the Brech, 43, 116; Rhine trench, 28, 31, 35, 36, 39, 114
Continental shelves: Atlantic Coast, 25, 50 (map), 90, 102; artesian conditions, 71, 74-81, 86 ff.; spring development on slope, 72-91; deltaic structure, 86; expulsion of non-artesian waters from, 89-91; sedimentation and consolidation, 90
Cooke, C. Wythe, vi, 76
Cooper, Margaret, vi
Coromandel coast: submarine springs, 73
Corsair Gorge, 13 f.

Cretaceous period, 75, 77, 82, 83, 85, 90, 100, 102, 107, 119; coastal plain, 10; deposits, 10, 53, 54, 55
Cronander, A. W., 46
Cuba: submarine spring, 72, 73
Currents: shallow water, 16; bottom, 24, 27, 36 ff., 42, 44, 46; density-current hypothesis, 24, 27, 28, 32, 39 ff.; hydraulic, 24, 26, 32, 46; littoral, 24; submarine, 24-61; tidal, 24, 25, 45, 46; types, 24-27; turbidity currents, 27-61; combined turbidity, salinity and temperature differences, 29, 31; reaction, 46, 57
Cushman, A. S., 97

Dalmatian karst: submarine springs off coast, 73
Daly, Reginald A., 8, 23, 27, 32-42 *passim*, 44, 46, 47, 52, 53, 54, 55, 56, 58, 114, 115, 116; quoted, 33, 35, 44
Dana, J. D., 19
Darton, N. H., 77
David, Sir Edgeworth, 73
Davis, William M.: turbidity-current hypothesis, 26, 31 f., 114; quoted, 100
Debris: evacuation of, 14 ff.; types deposited by muddy rivers, 49; canyon filling by, 104; *see also* Sediment; Silt
"Deeps" in ocean floor, 105, 106
Delaware: canyons in shelf margin off coast, 55
Delaware Bay, map, 50
Delebecque, A., 30, 38, 42; quoted, 40
Deltaic deposits, 47-51 *passim*, 86
Density: difference between Atlantic and Mediterranean waters, 45, 115
Density-current hypothesis, 24, 27, 28, 32, 39 ff., 114
Denudation of land surface by solution, 96 f.
Deposition, differential: as cause of river trenches, 29 ff., 40, 42; as cause of submarine canyons, 41, 43
Deposition of mineral matter: waters expelled by, 89 f.
Depth changes: accuracy of measurements of, 16
Diastrophism, 8, 11
Down-wearing of land, 83
Drowned valleys, 3, 8, 10; hypothesis, 8-9, 27, 113; re-excavated by landsliding, 9-18, 113
Dubalen, P. E., 69, 70, 73, 117
Du Parc, L., 39

Earthquakes: Japan, 15; submarine, 12 ff.
Eaton, H. N., 60
Eberhard, Graf von Zeppelin, *see* Zeppelin
Echo-sounding, 13, 14
Ekman, F. L., 24, 46
England: artesian water in sedimentary beds off coast, 85; solvent denudation, 96
Eocene formation: artesian conditions, 77
Erosion: headward, 18; hypothesis of origin of Rhine and Rhone trenches, 29 ff., 39, 115; conditions during early cycles of, 82; solution supplemented by, 101; mudflows as a cause of, 104
Evaporation: Mediterranean Sea, 45, 46

Fall Zone peneplane, 82
Faris, O. A., 60
Faulting, 4
Faults, submarine: solution along, 68, 69; non-deposition along, 69, 70
Feldspar, solubility, 94, 101
Fischer, Theobald, 45
Fjords, 3
Flint, R. F., vi
Florida: steep-head valleys, 91, 92 (fig.)
Forel, F. A.: quoted, 40, 42, 43; on Rhine and Rhone trenches, 29-44 *passim*, 114, 115
Foundering of subterranean caverns, 67, 117
French coast: subterranean river, 68
Fresh water: floating on sea water, 43, 51, 55-57; precipitation in salt water and, 47-52, 56, 100, 115; preventing deposition along fault, 69

Ganges River, 56
Gaskell, T. F., 85
Gay Head: clays, 53
Geneva Lake: Rhone trench, 28, 31, 35, 36, 37, 38, 39, 114; the Bataillière, 43, 116
Geologic past: artesian conditions, 82-85
Georges Bank: submarine canyons, 9, 15, 18 (map), 92; landsliding after earthquake, 13; submarine currents, 25-26; lithified sediments, 52, 53, 54, 55
Georgia: artesian wells, 78
Gibraltar: currents in Strait of, 44, 115
Glacial period: sea-level changes, 8, 10, 31 f., 82; canyon filling, 15; turbidity currents, 27, 31, 33, 114; storm waves, 58; sedimentation, 59

Glacial-water rivers: plunge under lake waters, 43
Gorceix, Ch., 67, 68, 70, 117
Gorges cut by spring sapping, 91
Göta-Elf outlet: reaction currents, 45, 46
Grabau, A. W., 86
Grand Canyon of the Colorado, 18, 92-93, 114
Great Plains: artesian system, 83
Gregory, J. W., 12
Groundwater sapping, 18-19, 91-93
Grover, N. C., 60
Gulf Stream: submarine valley, 77

Hadding, Assar, 103
Hall, C. M.: quoted, 91
Hall, James, 77
Harrisburg erosion cycle: artesian conditions before, 82
Hawaii: artesian conditions, 72; lapiés, 97
Hayes, C. W., 97
Headden, W. P., 95
Hedberg, H. D., 89
Heilprin, Angelo, 31
Heim, Albert, 30, 38, 39, 44
Helland-Hansen, Björn, 45
Hitchcock, C. H., 72, 73
Hofeneder, K., 73
Holt quadrangle, Florida, 91, 92
Hornblende: solubility of, 95, 96
Hörnlimann, J., 28, 114
Houston: artesian wells, 82
Howard, Arthur, vi, 77
Howard, C. S., 60
Hubbert, M. King, vi
Hudson River, map, 10; submarine trench, 19, 31, 54
Hugli River, 48
Hydraulic currents, 24, 26, 32, 46

Idaho: lava plateaus, 92, 97, 98
Insoluble: meaning of term, 94

Jack, R. L., 75
Japan: earthquake, 15; subterranean river, 66
Jardines: submarine springs, 73
Johnson, Douglas: quoted, 45, 71
Joly, J., 48, 100
Journal of Geomorphology, vi

Kattegat submarine channel, 24n
Kay, G. M., vi
Keilhack, Konrad, 73
King, F. H., 89, 90, 102

Kleinschmidt, E., 30, 38
Kuenen, Ph. H., 33, 59

Lakes: entry of rivers into, 28, 39, 42; fine materials held in suspension, 42; *see also* names, *e.g.* Geneva, Lake
Landslides: submerged river gorges re-excavated by, 9-18; evidence of, 12-18; submarine, 23-24; *see also* mudflowing
Lava plateaus: solution by springs in, 71, 91, 97, 98
Lawson, A. C., 86
Le Conte, Joseph, 72
Leman, Lake, *see* Geneva, Lake
Limestone: solution in, 97, 101; water expelled during consolidation, 89
Linhardt, Ernst, 31
Lithification of shelf sediments, 52-55
Littoral currents, 24

McCallie, S. W., 78
Mackin, J. H., vi
Malabar coast: submarine springs, 73
Martha's Vineyard: clay, 52, 53
Maryland: canyon off coast, 55
Maury, M. F., 45
Mead, Lake, 60
Mediterranean Sea and Atlantic Ocean: difference in waters, 44 ff.
Meinzer, O. E., vi, 82, 89; quoted, 81
Melton, F. A., v
Merrill, G. P., 97
Mill, H. R., 73
Milne, John, 12, 66, 73, 74; quoted, 74
Minerals: artesian waters, 78; waters expelled by deposition of, 89 f.; solubility, 94 ff., 119
Mississippi River, 47, 48, 56, 72, 86, 89
Missouri: artesian wells, 78
Missouri River, 48
"Mock valleys," 100
Monterey submarine canyon, 32
Mossom, Stuart, 76
Mudflowing, 11-17 *passim*, 23, 106; rôle of, in canyon evacuation, 103-5, 107, 119
Muds: water content, 89
Müller, R., 95

New Jersey, Appalachians and Coastal Plain, 84 (fig.)
Newport Pier: submarine canyon at, 16
New South Wales: artesian waters, 79
New Zealand: subterranean rivers, 66
Nile River, 50, 51, 56, 86

Niobrara formation: clay, 48
Non-artesian waters: expulsion from continental shelves, 89-91, 118
Non-deposition along faults, 69-70
North America: Coast, *see* Atlantic Coast
Notches: produced by spring sapping, 18, 92, 93

Ocala artesian aquifer, 76, 79
Ocean, turbid river waters floating on, 43, 56; effects of salinity and temperature on turbidity currents, 55-57
Ocean water, *see* Sea water
Oscillations of land and sea, 8, 93, 107, 113

Paleozoic period: canyon cutting, 9, 10; sediments, 90
Peedee deposits, 55
Pleistocene period, 10, 55, 82, 99, 119
Pliny, 43
Popovo polje, 74
Porosity of sediments, 89, 90
Port Macdonnell: submarine springs, 73
Precipitation: in salt and fresh water, 47-52, 56, 100, 115; by acids and alkalies, 48
Pressures at depth: aid to solution, 78, 95, 100, 119

Quartz: solubility of, 97, 98, 99
Queensland: submarine springs, 73; artesian conditions, 75

Ramsey, M. M., 73
Ravines: minor, 23, 103
Reaction currents, 46, 57
Reade, T. Mellard, 96
Reclus, Élisée, 73
Redondo Pier: submarine canyon at, 16, 17
Reservoirs: flow of turbid water through, 59-61
Rhine River: sublacustrine trenches, 28-31, 35-44, 114; the Brech, 43
Rhone River: sublacustrine trenches, 28-31, 35-44, 114; the Bataillière, 43; matter in suspension at mouth of, 51
River gorges: recently submerged, 8-9; ancient submerged, 9-18
Rivers: subaqueous trenches produced by entry into lakes, 28; factors other than turbidity in problem of river-lake densities, 39, 42; sediment-laden streams having no trenches, 42; sinking of heavier water under lighter lake water, 43; turbid water of, floating on ocean water, 43, 51, 55-57; bars or deltas at mouth, 47-51 *passim*; settlement of silts, 47-52; debris, types carried, 49
Rivers, subterranean: collapse of caverns formed by, 67; solution along faults, 68-69; non-deposition along faults due to up-rising waters, 69-70, 117; outlets, 66-67, 117
Roche-Poncié, M. J. de la, 66, 73
Rocks: in canyon walls, 52 ff.; solubility, 94 ff., 119; cementing value of powdered, 97
Rogers, W. B. and R. E.: studies of solubility, 95, 96
Rowan, A. S., 73
Russell, I. C., 92, 99; quoted, 98
Russell, R. J., 86

Sagami Bay: landslides and mudflows, 14, 15
St. Martin, Cape: submarine spring, 73
Salinity: effect on deposition of fine sediment, 47-52, 55-57, 115
Salinity currents: contrasts in, 24; Strait of Gibraltar, 45 ff.
Salisbury, R. D., 86
Salis-Soglio, Adolf von, 28, 35, 39, 41, 114; quoted, 36
Salt in solution: precipitating effects, 48
Sands, 98, 99; porosities, 89
Sandstone, 89; in canyon walls, 55
Sapping: groundwater, 18-19, 91-93; *see also* Spring Sapping
Sarasota County, Fla.: groundwater resources, 75-76
Schloesing, Ch., 42, 48, 50
Schmidle, W., 30, 39
Schneider, E. A., 48
Schooley erosion cycle: artesian conditions, 82, 83, 85
Schooley peneplane, 82
Schriever, William, v
Scott, W. B., 31
Sea level changes: Glacial time, 8, 10, 31 f., 82
Sea water: precipitation in fresh water and, 47-52, 56, 100, 115; contamination of wells, 76; access to artesian aquifer, 79; expulsion from continental shelves, 89-91; *see also* Salinity
Sediment: evacuated by landslides, 10; cables buried by, 12; density of currents charged with, 27, 114; effect of salinity

Index

on deposition of fine, 47-52, 116; degree of lithification of shelf, 52-55; behavior of subaërial and submarine, 54; deposition during pre-Glacial and Glacial times, 59; in reservoirs, 59; water incorporated in, and expelled from, 89-91, 103; *see also* Debris; Silt
Seelheim, F., 89
Sellards, E. H., 91
Shaler, N. S., 53
Shales, 89
Shand, S. J., vi
Sharp, Henry S., vi, 91
Shaw, E. W., 89, 90
Shelf-margin notches, *see* Submarine canyons
Shepard, F. P., 8-16 *passim*, 23, 25, 32, 55; quoted, 10
Sheyenne River: springs and coulees in delta, 91
Sidell, W. H., 47, 50, 51
Silicate minerals, 95, 97, 101
Silt: on floors of canyons, 26; in reservoirs, 60; on seaward face of a shelf, 87; water expelled during consolidation, 89; water content, 89, 103; varying results of deposition, 103; mudflow deposition, 103 ff.; canyon filling by deposition of, 104; *see also* Sediment; Debris
Sink-holes, 68, 69, 100
Skagerrak, strait: channel, 24*n*
Skey, William, 47, 50
Slano, Bay of: turbid water columns, 74
Slates, 89
Slumping, 87, 88, 93
Smith, J. Fred, 23, 33, 71, 104
Smith, Paul A., Jr., 32
Smith, W. S. T., quoted, 67
Snake River, 91, 98, 101
Sodium: carbonate, 78; chloride, 78, 81; sulphate, 79
Solubility of minerals in artesian waters, 78, 79
Soluble: minerals classed as, 94; term, 94
Solution: along faults, 68-69; rôle of, in spring sapping, 93-103, 119; studies of mineral and rock solubility, 95; of powdered rock, 97
Solution currents, 27, 114
Solvent denudation of land surface, 96 f.
Somerville erosion cycle: artesian conditions before, 82
Sorby, H. C., 89
Soundings, 3, 13, 14, 15

South America: cable breaks, 74
Springs, submarine: sapping by, 71-109, 118; in shallow and deep water, 72-74; development on continental slope, 72-91; competence of, to excavate canyons, 91-105; rôle of solution in sapping, 93-103; form and distribution of canyons compatible with spring-sapping hypothesis, 105-7
Spring sapping: above sea level, 18-19, 91-93, 98-99; submarine, 71-109, 118
Stearns, H. T., 92, 98, 99, 119
Steep-head valleys, 91, 92 (fig.)
Stephenson, L. W., vi
Stetson, Henry C., 23, 33, 53, 54, 55, 71, 92, 104; quoted, v, 25, 26, 52, 54
Storm waves beating upon continental shelf in Glacial time, 58
Stringfield, V. T., 73, 75, 79
Subaërial origin hypotheses, 8-22
Sublacustrine trenches of Rhine and Rhone rivers, 28-31, 35-44, 114
Submarine canyons: description, 3; genesis of forms, 3; hypotheses involving tectonic and non-tectonic origins, 4-6, 113; major and tributary, compared, 5, 113; subaërial origin, 8-22; maps, 10, 18, 50; continental shelves preceded by, 9; not always at river mouths or channel ends, 18; submarine origin, 23-65, 113; minor ravines not to be confounded with, 23, 106; silt deposit, 26; greatest depths, 41; theory of origin of sublacustrine trenches and, must differ, 44; subterranean origin, 66-112, 117; gradual production of fewer and larger, 102; whether filling by silt deposition and by coarser debris could occur, 104; forms classified apart from, 105; form and distribution of, compatible with spring-sapping hypothesis, 105-7
Submarine currents, 24-61; *see also* Currents
Submarine landslides: *see* Landslides
Submarine origin hypotheses, 23-65
Submarine spring sapping: hypothesis of canyon origin 71-109; hypothesis first stated, 71
Submerged gorges, *see* River gorges
Subterranean origin hypotheses, 66-112, 117
Suspension: fine material held in, 42, 114, 115
Suspension currents, *see* Turbidity currents

Tampa artesian aquifer, 75, 77, 79
Tectonic and non-tectonic origins: hypotheses involving, 4-6, 113
Temperatures: effects on submarine currents, 24, 28, 36 ff., 55-57, 114; artesian waters, 78; at great depths, 95, 100, 119
Temple, W. G., 45
Tertiary period: 82, 83, 85, 102; canyon cutting, 9, 10; deposits, 10, 53, 54
Thames River, 47; salts, 51
Thompson, D. G., 80
Tidal currents, 24, 25, 45, 46
Time necessary to produce canyons by spring sapping, 94-102, 107, 119
Torpats, Mrs. John, vi
Trenches: shallow, on continental shelf, 3, 8; of Rhine and Rhone rivers, 28-31, 35-44, 114
Turbidity currents: hypothesis, 27-61; trenches of Rhine and Rhone, 28-31, 35-44, 116; conclusions respecting hypothesis, 33-35, 61, 114; Strait of Gibraltar's currents, 44-46; effect of salinity on deposition of fine sediment, 47-52, 115; degree of lithification of shelf sediments, 52-55; effects of salinity and temperature on, 55-57, 114; inadequacy, 57-59; flow through reservoirs, 59-61
Tuscaloosa artesian aquifer, 100

Uhler, P. R., 53

Valleys: submarine canyons resemble, 8, 106; headward growth by spring sapping, 18-19, 91-93, 98-99
Veatch, A. C., maps, 10, 18, 50, 106
Vernon-Harcourt, L. F., 47 ff., 115; quoted, 51

Waldie, D., 48
Wales: solvent denudation, 96
Warping, 4
Water: amounts incorporated in, and expelled from, sediments, 89-91, 103; solvent and decomposing power of pure and carbonated, 94, 95, 119; meteoric, 96; *see also* Sea water
Wave agitation: mingling of waters by, 56, 59
Wells: contamination by sea water, 76
Wey, J., 30, 41
Weyquosque clay beds, Martha's Vineyard, 53
Wheeler, Girard, vi
Wheeler, Mrs. Girard, vi
Wheeler, W. H., 47 ff., 115; quoted, 51
Willard, D. E.: quoted, 91
Willis, Bailey, 97

Zeppelin, Eberhard, Graf von, 30, 38

Bei Fragen zur Produktsicherheit wenden Sie sich bitte an:
If you have any questions regarding product safety,
please contact:

Walter de Gruyter GmbH
Genthiner Straße 13
10785 Berlin
productsafety@degruyterbrill.com